Controlled Structures with Electromechanical and Fiber-Optical Sensors

Controlled Structures with Electromechanical and Fiber-Optical Sensors

Uri Melashvili,
Georgi Lagundaridze
and
Malkhaz Tsikarishvili

Nova Science Publishers, Inc.
New York

LIBRARY OF CONGRESS CATALOGING-IN-PUBLICATION DATA

Melashvili, Yuri.
Controlled structures with electromechanical and fiber-optical sensors / authors, Yuri Melashvili, Georgi Lagundaridze, and Malkhaz Tsikarishvili.
p. cm.
Includes index.
ISBN 978-1-60692-486-0 (hardcover)
1. Structural control (Engineering)--Equipment and supplies. 2. Optical fiber detectors. 3. Electromechanical devices. 4. Flexible structures. 5. Smart structures. I. Lagundaridze, Georgi. II. Tsikarishvili, Malkhaz. III. Title.
TA654.9.M465 2009
624.1'71--dc22
2008048994

Published by Nova Science Publishers, Inc. ✝ New York

CONTENTS

PREFACE

The content of the book is continuation of the previous book, it was published in Poland in 2001, co-autor prof. Andrzej Flaga it refers to regulation of voltages and deformations in combined structures (in constructions with cables or guy ropes), this book is provided with our copyright. As the sensor of deformation the spring was used and were reviewed beam (cantilever and double-support) combined construction.

Now in the given book framed and spatial combined structures with application electromechanical and fiber-optic sensors are considered (Geopatent P2728 and USSR patent 54009). The control behavior of constructions at all phases of their life cycle and prior preventive alarm without participation of the person is an actual problem and considerable achivement of technical development and new technologies.

The book will be interesting for the desingners, tachers of University, bachelors and holders of masters's degree as well as for young and skilled scientists.

ACKNOWLEDGEMENTS

We would like to thank all those who have assisted in various ways in the preparation of this book.

We are especially grateful to Full Professor Dt.Tech. science Revaz Tchvedadze, for examination some date for his publications. Lia Balanchivadze and Gina Gureshidze for calculations some problems and to Ms. Tinatin Magradze and Ms. Tamar Skhiladze for typing and drawings the manuscript, Nani Tsenteradze for translating the manuscript.

The authors has had advice in particular from Mr. Sergo Gotsiridze who made helpful suggestions.

UNITS AND CONVERSION FACTORS

UNITS

While most expressions and equations used this book are arranged so that they are non-dimensional, there is number of exceptions. In all of these, SI units are used which are derived from the basic units of kilogram (kg) for mass, metre (m) for length, and second (s) for time.

The SI unit of force is the newton (N), which is the force which causes a mass of 1 kg to have an acceleration of 1 m/s^2. The acceleration due to gravity is 9,807 m/s^2 approximately, and so the weight of mass of 1 kg is 9,807 N.

The SI unit of of stress is the pascal (Pa), which is the average stress exerted by a force of 1 N on an area of 1 m^2. The pascal is too small to be convenient in structural engineering, and it is common practice to use eitner the megapascal (MPa=10^{6}Pa) or the identical per square millimetre (N/mm^2=10^{6}Pa). The megapascal (MPa) is used throughout this book.

le of Conversion Factors

To Imperial (British) units	To SI inits
1 kg = 0.068 53 slug	1 slug = 14,59 kg
1 m = 3,281 ft	1 ft = 0,304 8 m
= 39,37 in	1 in = 0,025 4 m
1 mm = 0,003 281 ft	1 ft = 304,8 mm
= 0,039 37 in	1 in = 25,4 mm
1 N = 0,224 8 lb	1 lb = 4,448 N
1 kN = 0,224 8 kip	1 kpi = 4,448 kN
= 0,100 36 ton	1 ton = 9,964 kN
1 MPa* = 0,145 0 kip/in^2	1 kip/in^2 = 6,895 MPa
= 0,064 75 ton/in^2	1 ton/in^2 = 15,44 MPa
1 KNm = 0,737 6 kip ft	1 kip ft = 1,356 kNm
= 0,329 3 ton ft	1 ton ft = 3,037 kNm

* 1 MPa = 1 N/mm^2

† There are some dimensionally inconsistent equations used this book which arise because a numerical value (in MPa) is substitutes for the Young's modullus of elasticity E while the yield stress F_Y remains algebraic. The value of the yield stress F_Y used in these equations should therefore be expressed in MPa, while care should be used in cinverting these equations to Imperial inits.

** Decanewton = 1 daN = 10 newton = 1 kg

INTRODUCTION

The modern achievements in the field of robotics and technology of bulding enable creations of buldings and with applications of which one are capable to adapt in a broken operation conditions.

In many buldings and facilities some kinds (views) of automatic devices of diffezent assidning such as, for example, automatically opening and accluded doors an automatic vobce circuit, an automatically monitoring telesystem, the automatic security signalling. System scroll bars with the automatic equipment, the automatic fire-prevention device, automatic telescopic bridges and locks an so on already operate.

Here opportunely to recollect about such phenomena as earthqauakes and hurricanes, temperature and radiation effects by which one the designs of buldings and facilities are subject.

Therefore it is possible to transmit a part of a protective fuction from these phenomena to automatic system as it is made in aircruft manufacturing.

Already tody in it there is a necessity in hard – to – reach and dangerous for the person places of building and facilities a case of necessiry of regulation by deflection stained of a construction and part this function to transmit automatic systems, example, for nuclear stations and space systems.

In the tendered book some engineering pathes for implementation of this purposes on examples already of activities executed as are intended.

Chapter 1

USE OF ELECTROMECHANICAL AND FIBER-OPTICAL SENSORS IN CIVIL ENGINEERING

1.1. ELECTROMECHANICAL SENSORS

Let's consider two versions of electromechanical sensors.

The first alternative is used in tower antenna systems in case of antenna raising and lowering. Here are used: a hoister with electric motor and cable or guy tension meter, for example, EMIN and PIN devices with sensors.

Using the sensors which transmit signal to hoist electric motor for its actuating when design stress in cable or guy exceeds its design value by ±10%. In this case it is possible that strain of antenna be kept unchanged with hoist and sensor devices.

The devices EMIN and PIN supplying signal to electric hoist consist of body having elastic element with sensors and electric block supplying signal to electric hoist (Figure 1.1).

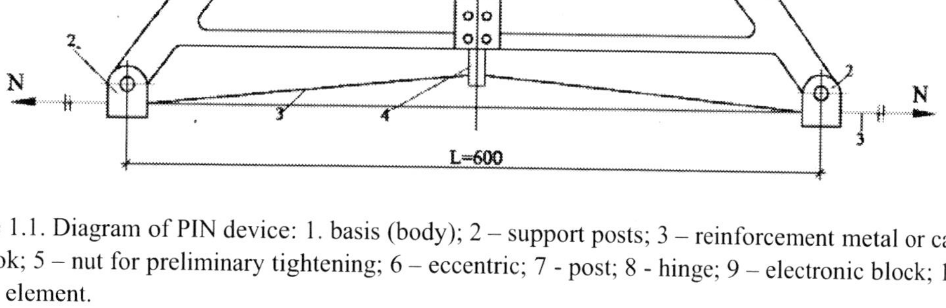

Figure 1.1. Diagram of PIN device: 1. basis (body); 2 – support posts; 3 – reinforcement metal or cable; 4 - hook; 5 – nut for preliminary tightening; 6 – eccentric; 7 - post; 8 - hinge; 9 – electronic block; 10 – elastic element.

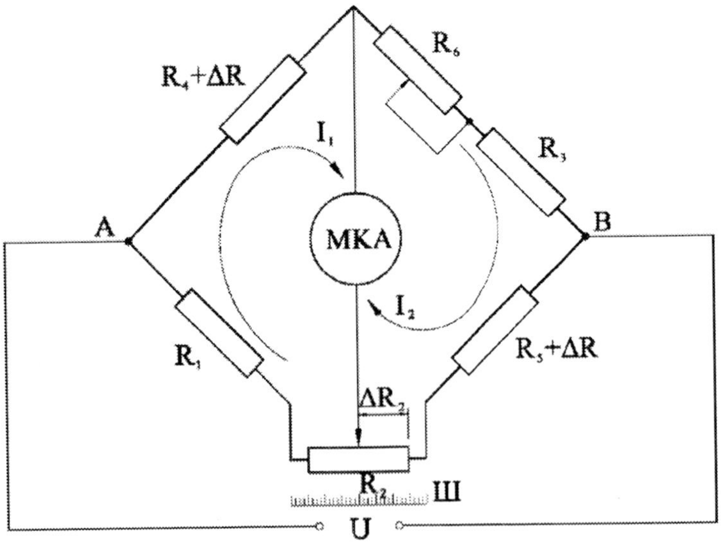

Figure 1.2. Diagram of balance bridge: U - power source; MKA – microampere meter; I_1, I_2 - currents in circuits; R_4+ΔR and R_5+ΔR – resistance of active sensors: R_1 and R_3 – resistance of compensating sensors; R_2 and ΔR_2 – resistance of rheochord; R_3 - fitting resistance; Ш – scale of rheochord.

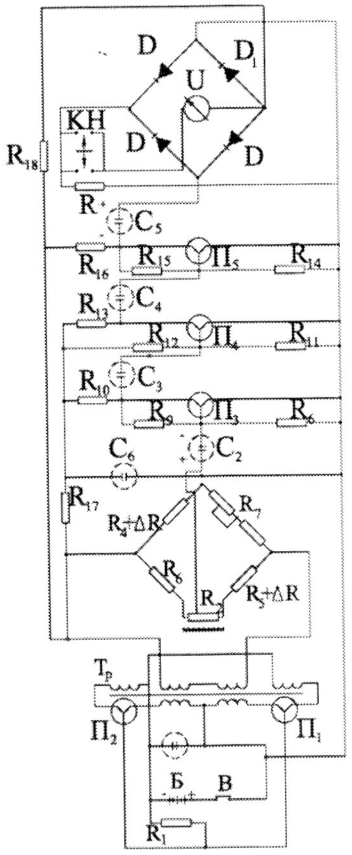

Figure 1.3. Schematic electric diagram of device EMIN-2.

Balance bridge connected to transmitters is contained in electronic block. The diagram of balance bridge is given in Figure 1.2.

Schematic diagram of the device is given in Figure 1.3.

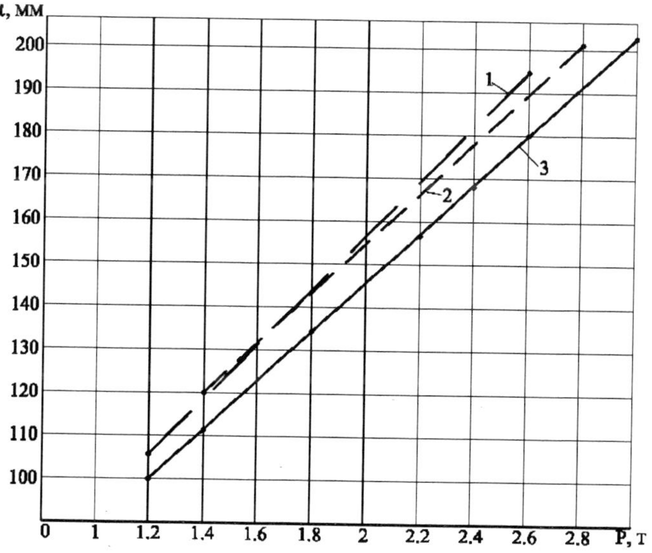

Figure 1.4. Calibration characteristics of device EMIN-2 for the wire and strands of diameter up to 6 mm α=*f*(P) 1- high-strength wire of class Bp11, ∅ 5 mm; 2 – the same, ∅ 6 mm; 3 - strand of П-7 class, ∅6 mm.

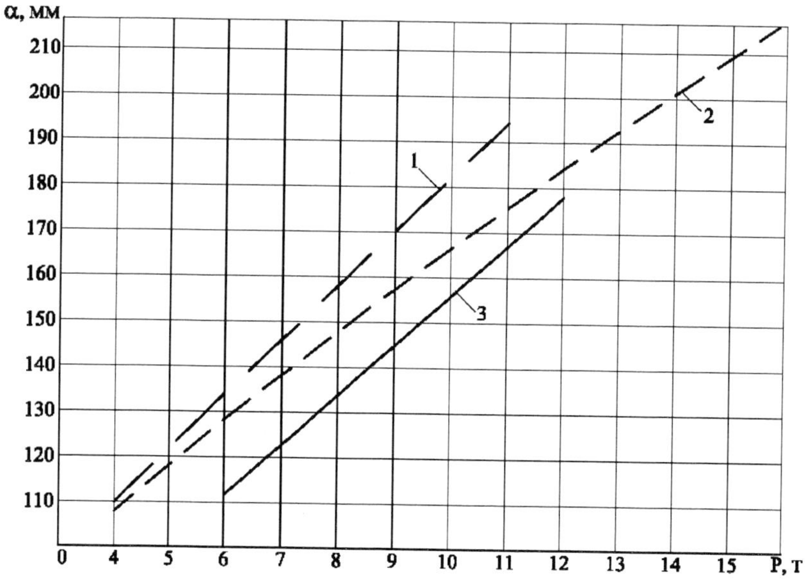

Figure 1.5. Calibration characteristics of device EMIN-2 for stem and strand reinforcement of average diameter α=*f*(P) 1 – stem-reinforcement of class A-IV, ∅12 mm; 2 – strand reinforcement П-7 ∅15mm; 3 – the same, ∅12mm.

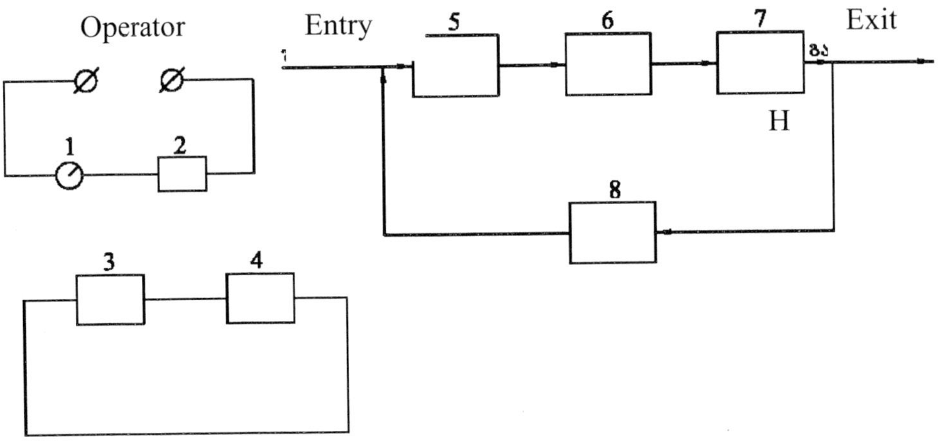

Figure 1.6. Schematic diagram of strain regulation.
1- dynamometer; 2 - tensioning device; 3 - turn-buckle; 4 - item; 5 - relay; 6 - actuating mechanism; 7 - item; 8 - dynamometer with contactors.

Micro ampermeter of the device registers the signal from the sensor the hand of which in ultimate case touches the contactor, circle closes and electric hoist is actuated.

The devices have gauging characteristics for different diameter and length cables and guys. Some gauging characteristics are given in Figures 1.4 and 1.5. The diagrams show the relation between strain force and rheochord deviation.

The second electro mechanical device which is also used for regulation of cable or guy tension is based on USSR author's right N 543609 [ ]. The mentioned device consists of electric motor with planetary screw transmission and relay spring dynamometer with contactors.

In the case of cable expansion and contraction (at design strain deviation), the cam placed on spring catches right or left contactors and the circuit is closed; respectively, electro motor rod advances and decreases cable or guy strain or goes back and increases the strain.

Schematic diagram for manual or automatic control of strain is given in Figure 1.6.

1.1.1. Regulation of Forces and Strains in Constructions with Electric Hoist

For regulation of forces and strains in constructions electric hoister is considered as a complex mechanism, the main elements of which are a drum and a cog-wheel transferring rotation from drive shaft to round shaft.

Electric hoists are used for tightening of semi-mechanized cables and guys.

Electric hoist consists of a frame on which electric motor connected to elastic coupler from cylindrical reducer is mounted. On the end of low-speed shaft a gear is mounted which transmits rotation to tooth gear and cylinder. Cable is coiled on the cylinder in four or five layers. Brake is an electromagnetic shoe, the second brake is a conveyer-type. The characteristics of electric hoists are given in Table 1.1.

Table 1.1.

N	Load capacity	Cable capacity, m	Number of coil layers, n	Cable diameter, mm	winding rate, m/min	Dimensions, mm			Motor capacity, kVt	Mass in kg without cable	Note
						Length	Width	Height			
1	1.0	60	3.0	13.5	13.0	1100	920	615	3.0	161.0	Cable GOST-3079-69
2	1.5	150	4.0	13.5	12.0	1350	1034	702	4.5	634.0	R breaking=1,666GPa
3	3.0	200	4.0	17.0	9.0	1636	1334	703	7.0	985.0	
4	5.0	315	5.0	21.5	5.0	1703	1620	1060	7.0	1530.0	
5	7.5	350	5.0	29.0	7.0	2250	1625	1277	10.0	2235.0	
6	12.5	800	7.0	33.0	7.65	2960	2310	1800	20.0	5580.0	

At cable winding around the cylinder the minimum diameter of the cylinder determined when cable bending behavior is practically excluded.

$D \geq d \cdot e$

where d is cable diameter in mm; e is the coefficient depended on mechanism type and on its behavior mode (see Table 1.2)

At hoist operation and in the case they are chooen it becomes necessary to define main technical characteristics and to calculate their fixation elements.

Cable capacity given in the Table (in meters) is determined with formula:

$$L_{cabl} = \frac{\pi z n}{1000}\left(D_{cyl} + d_{cabl} n\right) - 2\pi D_{cyl} / 1000$$

Here Z is the number of cable coils on the cylinder when coil lead is $t=1{,}1d_{cabl}$;

$z = \frac{L_{cyl}}{t}$ where L_{cyl} is working length of the cylinder;

n is the number of cable coil layers on the cylinder;

D_{cyl} is cylinder diameter, mm;

d_{cabl} is cable diameter, mm.

Hoist electric motor required power (kVt) is determined with formula:

$$N = SV_{cab} \Big/ (102 \cdot 60\ \eta)$$

Table 1.2. Minimum allowed value of coefficient e

Type of mechanism	Actuator and its behavior mode	Coefficient e
Hoist	Manual	12.0
	Machine	20.0

where S is hoist tractive force (kgf or kN);

V_{cab} is the rate of cable winding on hoist cylinder (m/min)

η is hoist efficiency which considers losses on cylinder and actuator:

$\eta=\eta_{cye}\cdot\eta_{act}$

Here η_{cyl} - cylinder efficiency for sliding bearing is equal to 0.96; swing of bearing equals to 0.98.

η_{act} - actuator efficiency for one pair of pinions the value of which is determined from the Table 1.3.

For n pair bearings the efficiency of actuator will be:

$\eta_n= \eta_1\cdot \eta_2\cdot\ldots \eta_n$

Mass of counterweight (in tons) which provides hoist stability from overtopping is determined with formula:

$$G_{counterweight} = K_y\left(sh - G_{hoist}\cdot \ell_2\right)/\ell_1$$

where K_y - stability coefficient equals 2.0;

s - is hoist tractive force (tf);

h - is cable height from ground level (m);

G_{hoist} - hoist weight (t)

ℓ_1 and ℓ_2 - distance from overtopping ridge to counterweight and hoist gravity center (m).

Horizontal displacement of hoist is done according to formula:

$$p = S - T = S - (G_{ho} + G_{cw})\cdot f$$

Here P is the force resisting to horizontal displacement of hoist (tf);

S is hoist tractive force (tf);

T hoist frame friction force on support surface (tf):

$$T_{fr} = (G_{ho} + G_{cw})\cdot f$$

Here G_{hoist} is hoist mass (t);

G_{cw} is counterweight mass (t);

f is friction coefficient at steel sliding on dry pebble f=0,45

Table 1.3. The values of hoist actuator efficiency η_{act}

Type of cylindrical gear	Efficiency of bearing η_{act}	
	Sliding	Swinging
With open milling bearings	0.93	0.95
With bearings milled in oil bath	0.95	0.97
With bearings polished in oil bath	0.98	0.99

Table 1.4. Technical characteristics of mounting electric hoist

Type of hoist	Tractive force,	Cable winding rate, m/min	Cylinder diameter, mm	Cylinder length, mm	Number of cable winding layers	Cable capacity	Electric motor power, kVt	Hoist mass, kg
Л-1001	1,0	23,0	168	470	3	75	4,5	287
Л-3003	2,0	8,5-17,5	299	-	-	600	7,2	1040
ЛТ-2500	2,5	21,6	400	875	1	40	7,5	1166
ЛЗ-50	3,0	42,0	300	800	5	260	16,0	1425
ПЛ-5-50	5,0	40	246	1160	4	450	22,0	1861
СЛ-5	5,0	6,2-42,7	-	-	6	1200	30,0	5100
ЛС-5-30-90	5,0	1,12-41,4	426	-	-	900	22,0	2440
ИЗ-587	7,5	7,0	500	-	-	350	10,0	2240
ЛТ-2М-80	8,5	35-48	-	-	-	1440	-	15500
ЛМЗ-10-50	10,0	10,6	-	-	5	510	-	3790
ЛМН-12	12,5	7,66	750	-	7	800	20	5643
Л-15-А	15,0	10,0	620	2400	4	600	30	8000
ЛМ-16/2500	16,0	5,7-8,6	800	2000	-	1250	32	10350
ЛМ-33/2000	32,0	9,0	-	-	-	2000	-	-

For the obtained P force the support elements are calculated against horizontal displacement of hoist.

The existing types of electric hoists are given in Table 1.4.

1.1.2. Actuator with Device by Epicyclic Gear Train

As it has been mentioned cylindrical gear is used in hoists. The hoist of the second type which was used in МП 100 has a planetary gear.

The advantage of planetary gear is its small mass (is 2÷4 times less than that of other gears), a greater number of transmissions (to thousand and more), small loads on supports that simplifies support construction but needs high precision of manufacturing and mounting.

1.2. Electromechanical Sensors for Building Structures Control

Even today we can outline some real ways and make a brief survey of already executed examples of the control of buildings and constructions.

For antenna structures counterweights are used depending on load in order to provide strain. In the absence of counterweight and its replacement with a sensor giving signal to hoist motor the reeving of cable happens in automatic mode (Figure 1.7).

If wire antenna is isolated from supports then hoists can be installed on the ground.

The Sinkansen highway in Japan is the transport system of Japanese islands, undergoing frequent earthquakes, therefore its structures are to have ample seismic resistance, as well as the means of automatic stopping of trains at violent earthquakes, that immediately react to the signal supplied from seismometers located in definite intervals on the expressway when seismic signal exceeds the definite level (Figure 1.8).

In the future it is supposed to use in cosmos large lattice constructions of great length and small mass. Assembly of grid rods, platforms and antennas of unprecedented dimensions is proposed.

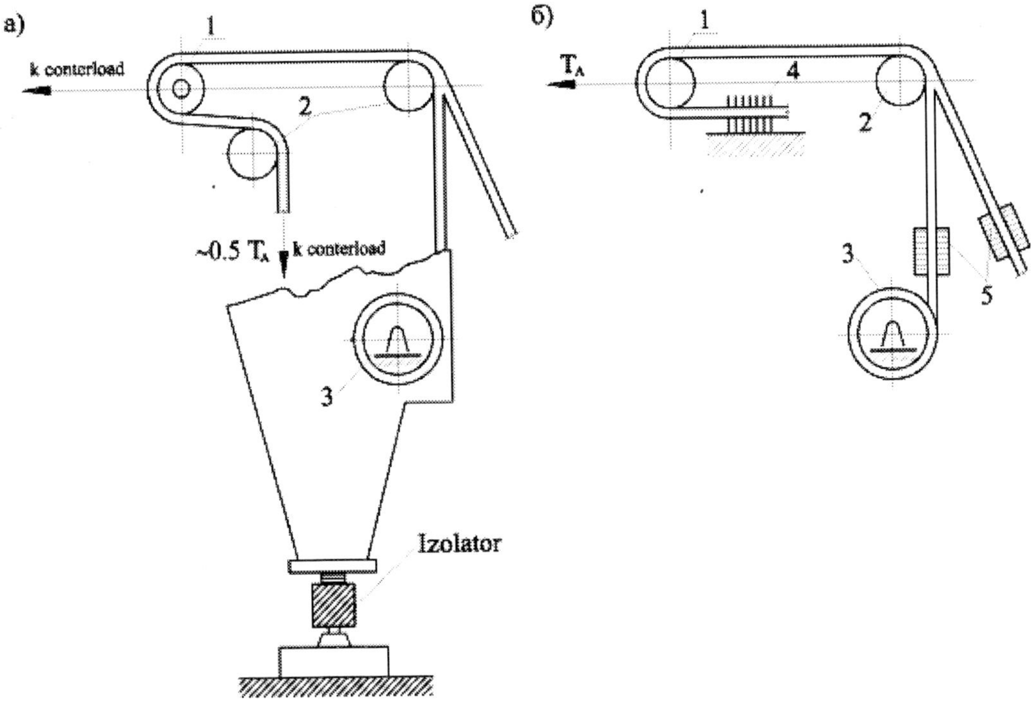

Figure 1.7. Diagrams of cables reeving running to the hoist.
a – in counterweights; b – case sensors; 1 – movable roller; 2 – immovable rollers; 3 - hoist; 4 – immovable bracing; 5 – sensor: in dotted line – of mounting of hoists on the ground without grounding.

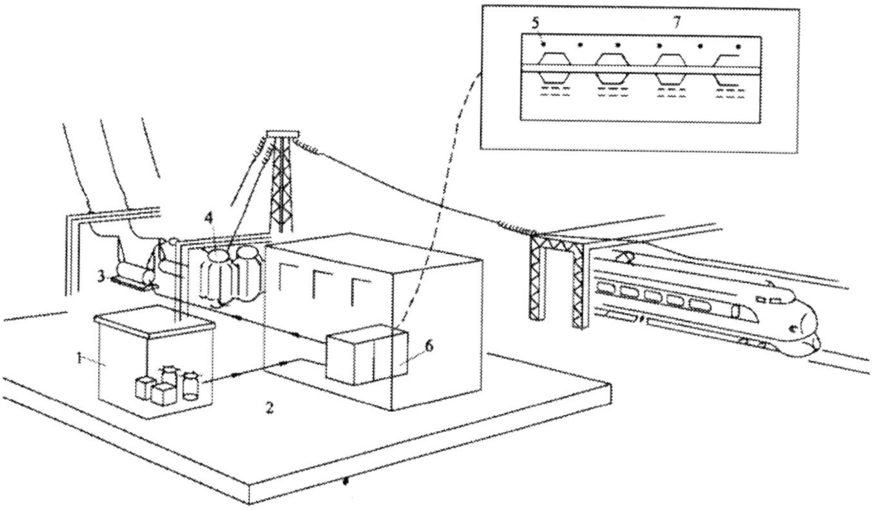

Figure 1.8. Diagram of protection system in case of earthquake.
1 – seismograph; 2 – traction substation; 3 – disconnector; 4 – transformer; 5 – signal device; 6 – control panel; 7 – central station of warning signals feed.

That fact that the loads acting on space structures are very small and structure dimensions are very great will indispensably be the reason of the elasticity of their members and will require stretched elements for provision of their strength and rigidity, particularly for great dimension reflectors and solar reflectors (Figure 1.9, 1.10).

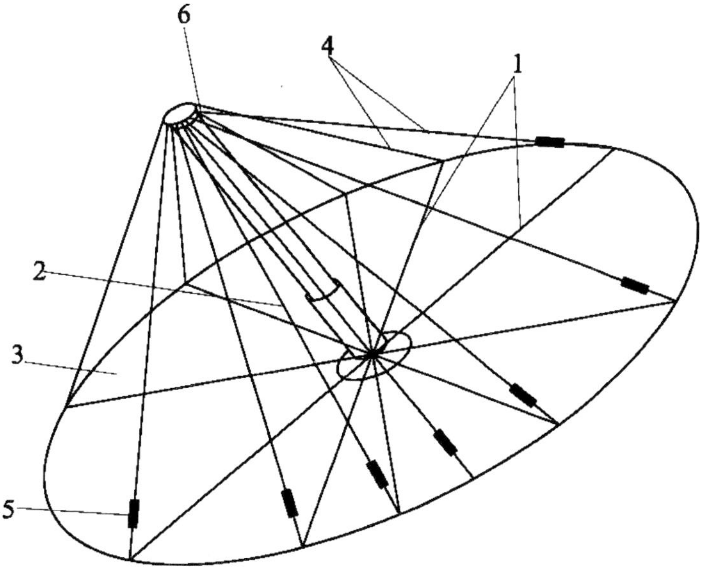

Figure 1.9. Space reflector. 1. – rigidity ribs; 2 – centering boss; 3 – reflector; 4 - guy-ropes; 5 – control sensors; 6 – telescopic rod.

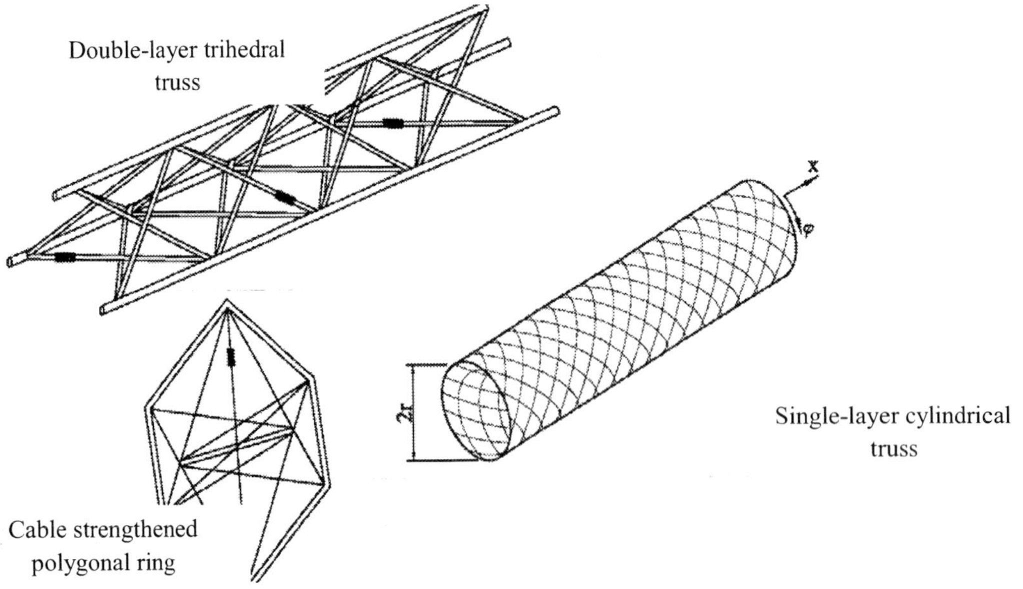

Figure 1.10. Space lattice constructions.

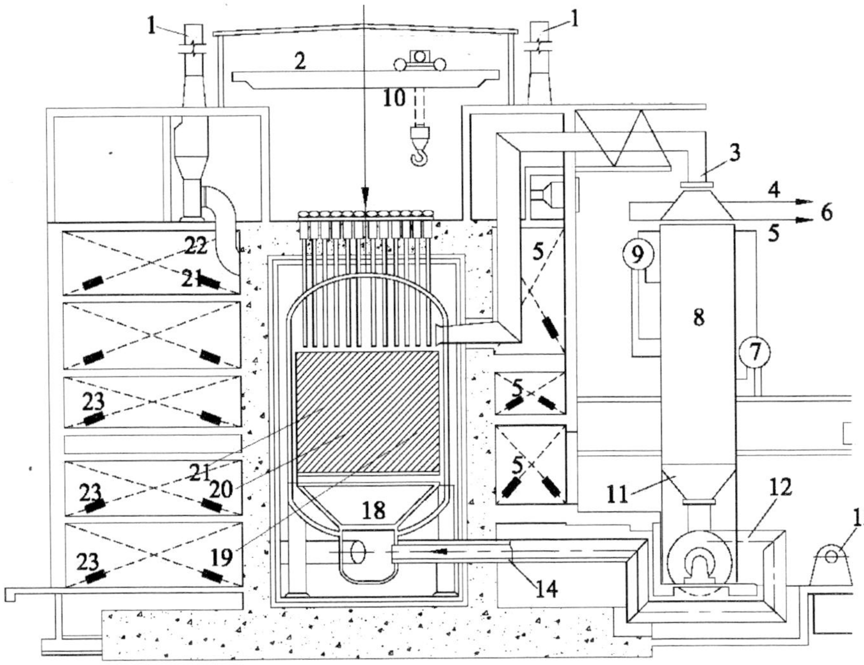

Figure 1.11. Cross-cut of power reactor building of atomic power station in Colder-Hall (England): 1 – gas cooling pipe; 2 – bridge crane; 3- hot gas; 4 – high pressure vapor; 5 – low pressure vapor; 6 – vapor to standard vapor turbogenerator devices; 7 – low pressure vapor catcher; 8 – heat exchanger; 9 – high pressure vapor catcher; 10 – bridge crane; 11 – cool gas; 12 – air blower of circulating cooling heat transfer agent and with electric drive; 13 – motor-generator; 14 – cool gas canal; 15 – heat protection; 16 – hot gas canal; 17 – loading tubes; 18 – eeservior under pressure; 19 – graphite; 20 – uranium bars; 21 – control bar; 22 – biological protection; 23 – control sensors.

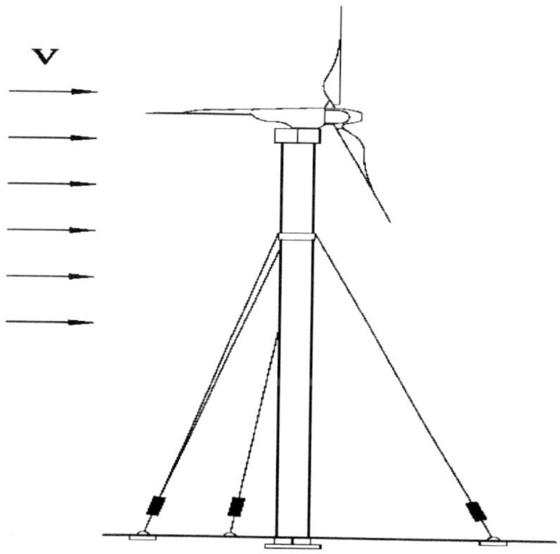

Figure 1.12. Power station masts.

The problem of providing seismic resistance of important power structures, constructions, equipment and pipelines of atomic power stations (APC) in seismically active regions is particularly actual as the failure of elements and constructions of atomic power station may lead to ejection of radiation substances on great territories with severe ecologic consequences.

At present the problem of equipment of buildings, constructions, devices and pipelines of muclear power plants (npp) and atomic power stations with programmable elements: guys, braces, etc. which in the process of seismic or other outer action control the oscillation mechanism, the adaptive systems are actuated that allow insignificant oscillations caused by technological processes and temperature displacements, but are actuated at setting threshold values of speed and accelerations at seismic action.

Adaptive systems can be placed in frame system of APS buildings, as well as, around reactor and other facilities (Figure 1.11)

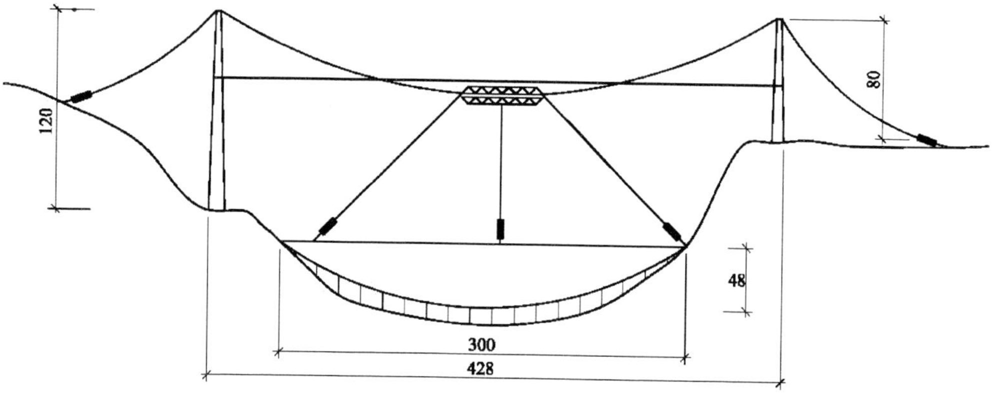

Figure 1.13. Radio telescope with guy stays.

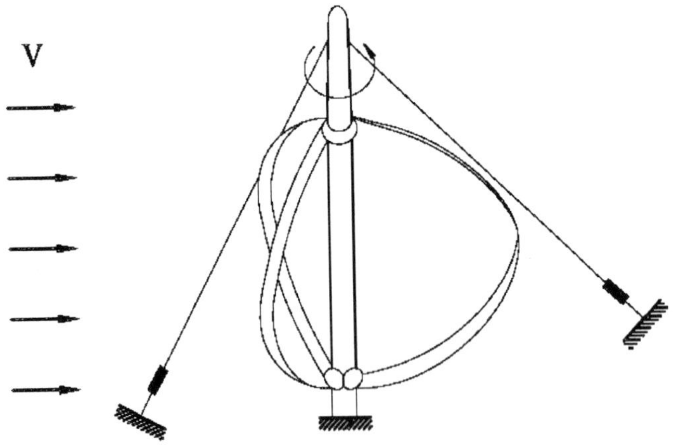

Figure 1.14. Wind power station masts with stays.

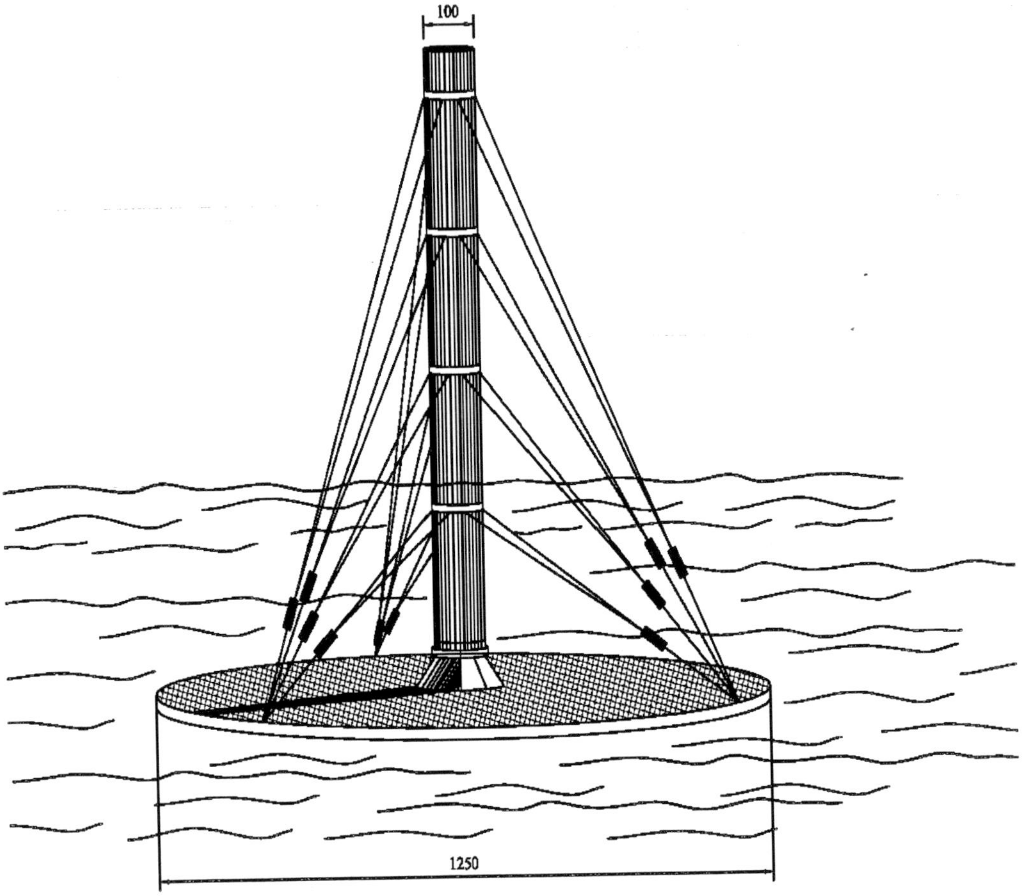

Figure 1.15. Solar power station tube with stays.

Many high-rise structures, such as towers, masts, tower buildings have guy ties which in some cases are already equipped with control sensors, as for example, mast antennas (Figure 1.12).

Particularly important is the system of automatic control of overground radio telescopes (Figure 1.13), also of wind and solar power stations (Figure 1.14 and 1.15), sea platforms and oil containers (Figure 1.16 and 1.17), composite large span coverings of buildings and structures, suspended and guy bridges, etc. (Figure 1.18). Possive and active vibro dampers (Figure 1.19).

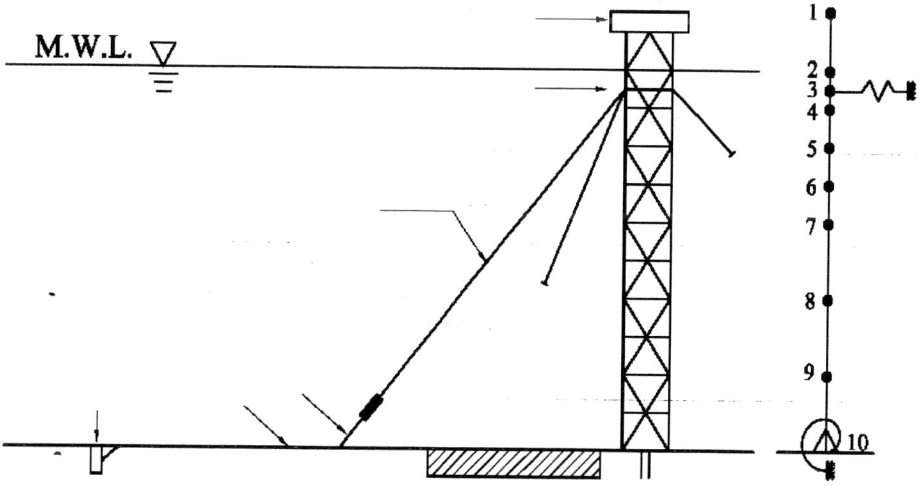

Figure 1.16. Structural configuration of offshore guyed tower.

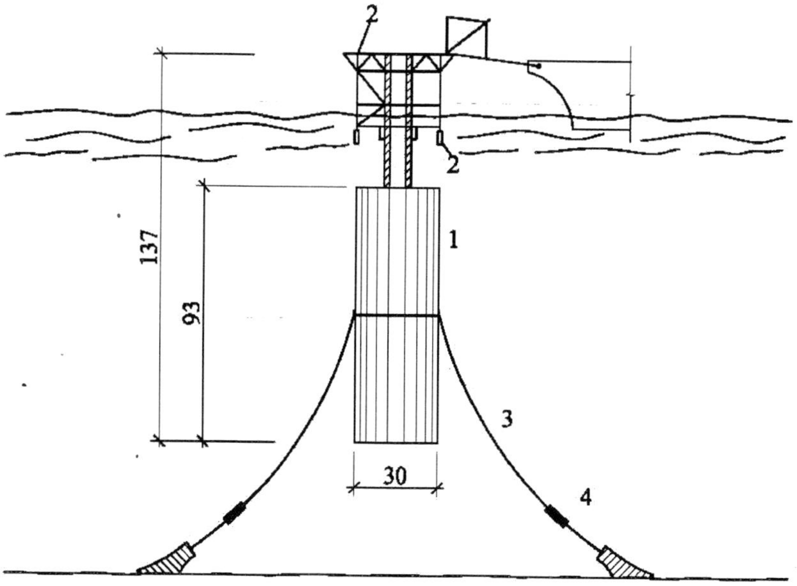

Figure 1.17. Submerged oil container.
1 – cylindrical reservoir; 2 – platform with pump station; 3 – anchor cable; 4 – control coupling.

Deck fairlead guy line clump weights anchor line anchor line a. elevation view b. structural model.

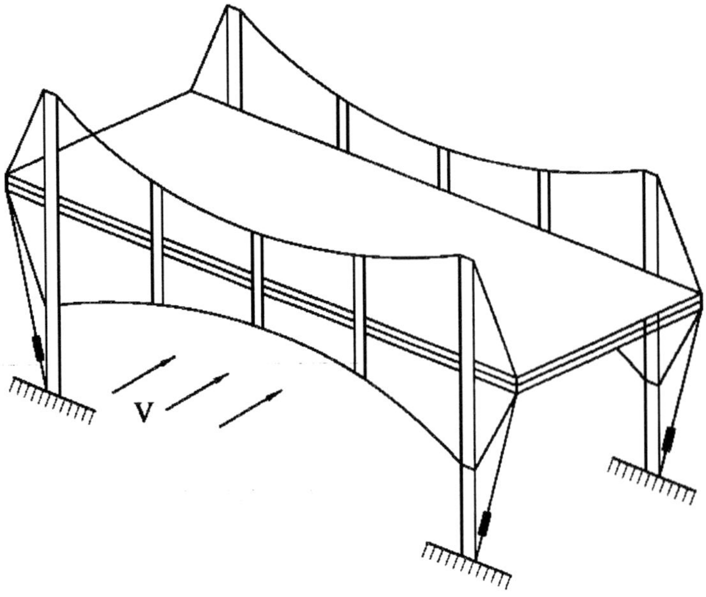

Figure 1.18. Composed guy covers and bridges.

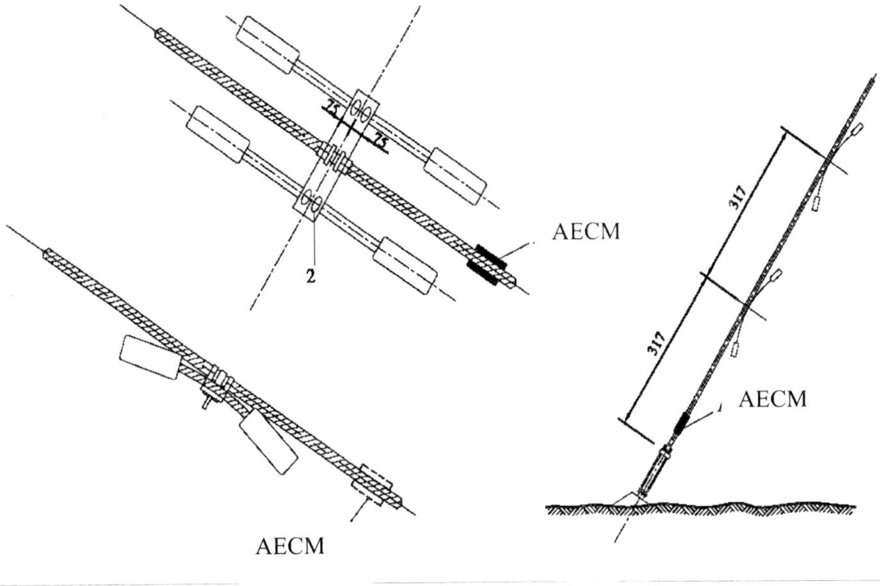

Figure 1.19. Passive and active vibro dampers.
a – vibro damper stays to cable; b – device scheme.

At present there are a number of monographs and works dedicated to controled structures, as for example: group of authors under the editorship of C..M.Belotserkovski, monographs by T.N.Soong, A.V.Perelmuter and N.P.Abovski, collection of reports under editorship of Kh.P.Laipkholtz.

One may agree with the author's statement that "the concept of active control of the behavior of bearing has a great future and even now the time has come to attentively consider and discuss the state of the problem" from different points of view.

As to transformable structures they, apparently should be considered separately from controlled structures as is proposed in monographs.

1.3. The Foundations of Creation of Optoelectronics and Fiber Optics

The end of the XIX century may be considered as the beginning of metrology. At that time a definite systematization has been done in the sphere of electro technique on the basis of alternating current and theory of electric magnetism. Until then physical values were mainly measured by mechanical means.

In the first half of the XX century electric measuring devices have gain spread, the operation principle of them being based on interaction forces of magnetic field and electric current. These devices were introduced in modern technologies. Electric measuring devices became the core of measuring industry and metrology.

The development of electronics in the second half of the last century caused colossal changes in metrology. The oscillograph has been created which contained some ten hundreds and more electronic bulbs. It had very high functional possibilities. Also a number of similar devices have been developed which gained wide application in scientific researches and industrial sphere. This was the beginning of the era of electronic measurements. At present elementary basis of measuring devices has been significantly changed. Electronic bulbs were substituted with transistors, chips, big integrated circuits. Electronics became the basis of measuring technique.

There is a great difference between electronic measurements of the 50th and electronic measurements of the 80th of XX century. The difference is that digital technique has been introduced into measuring devices.

The necessity of signal processing with different electronic circuits is continuously arising in measuring system. In the development of digital measuring technique it is surmised that digital signal is fed directly from sensing element of the sensor. In most cases this signal has analog form and on the entrance of data processing block there is mounted analog-digital transformer. Digital technique is mainly used in data processing block and in output device (indicator) or in one of them.

The main advantage of digital technique is comparatively simple realization of high level operations that are difficult to realize with analog devices. The measurement of quite small values becomes possible.

The precondition of development of fiber-optical sensors was the functional expansion of the operation performed in data processing block of sensor, by their numbering and simplification of nonlinear type operations.

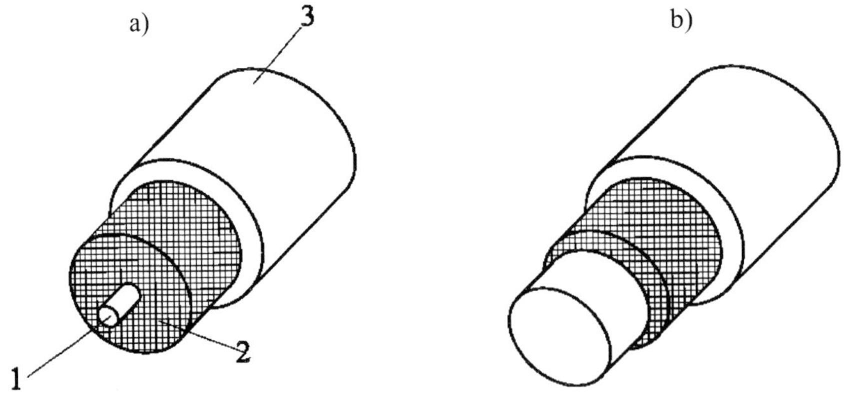

Figure 1.20. Optical fiber diagram.
a - dingle mode; b - multi mode a- core; 2 - jacket; 3 – protecting layer.

The main stimulus of fiber-optical sensors production was the creation of optical fiber. This will be discussed below.

On the basis of optics and electronics a new sphere of science and technique – optoelectronics - has been developed. 1950 may be called the year of birth of optoelectronics when E.Loebner described potential parameters, so-called optrons, of different optoelectronic devices of connection. After then optoelectronics has been continuously developed. The invention of lasers in 1957-1960 stipulated the acceleration of optoelectronics development.

The basic moment in the development of optoelectronics is the creation of optical fiber. Especially intensive researches were begun in the 60th of the past century while in the 70th American firm “Korning” began fiber production from quartz (20 dm/km), that stimulated other researches to work for improvement. At present 2 dm.km and less losses are achieved.

The main purpose of optical fiber creation was its using in communication lines. In this direction very great success is achieved, almost evey where in the world the fiber successfully substituted copper wire. It is used in internet circuits and in computer-informational technologies. At the end it was decided to use its unique properties in fiber-optical sensors which is developing very successfully.

Optical fiber, as is shown in Figure 1.20, is a transparent glass cylinder consisting of core and jacket. Light is propagating in the core. Optical fiber may have outer layer – cover - that protects fiber surface, increases its strength and simplifies its exploitation.

Refractive index of core n_1 is negligibly more than refractive index of jacket n_2, therefore, light fed to core is reflected from core-jacket border and is entireby propagated in the core (Figure 1.21). For full-scale internal reflection the aperture angle of light fed into core is equal to:

$$Q_{maq} = \sqrt{n_1^1 - n_2^2} \tag{1.1}$$

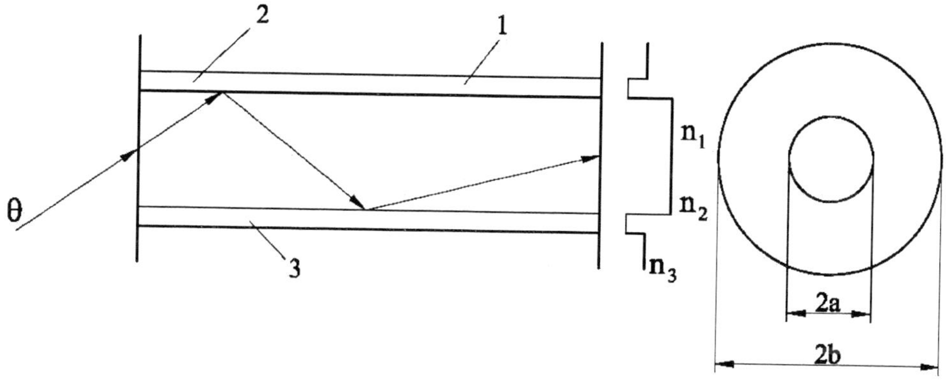

Figure 1.21. Typical structure of optical fiber and light propagation.
1 – core; 2 – jacket; 3 – protection layer.

Out of physical properties those properties of optical fiber are more frequently considered which are more important at using as a sensitive element, more exactly, such as the character of light phase change propagated along the fiber - variation under the effect of mechanical pressure, temperature, magnetic and electric fields. Besides, it is necessary to consider characteristics connected with radioactive radiation.

According phase variction it is possible to crate a relatively sensitive sensor for pressure, temperature, magnetic, electric and other measurements.

Generally there exist two types of optical fibers: single mode one where only one mode is propagated (transmitted electromagnetic field distribution type) and multi mode – with transmission of many modes (about hundred). Constructionally these types of fiber differ only in core diameter – light conducting part, within which refractive index is slightly more than in jacket (Figure 1.21).

In technique multi mode as well as single mode optical fibers are used. Multi mode fibers have a great diameter of core (about 50 mkm), that simplifies their interconnection. But as group rate of light is different for each mode at transition narrow light pulse it widens (increase of dispersion). Compared to multi mode fiber the advantages and disadvantages of single mode fiber interchange: dispersion is reduced but small diameter of the core (5…10 mkm) comparatively complicates the connection of this type of fibers and laser light beam introduction into them.

As a result single mode optical fiber has gained advantageous application in communication lines requiring high speed of information transmission while multi mode ones more often are used in communication lines needing relatively low speed of information transmission.

In fiber-optical sensors except those of interferential type where single mode optical fibers are used multi mode optical fibers are everywhere used that simplifies the creation of metering systems.

Consider optical fiber characteristics for communication systems, as well as, for structural element of a transmitter.

At first note general advantages of optical fiber:wide bandedness (up to tens of meganertz), small losses (minimum 0.154 .km); small diameter (about 125 mkm); small mass; (about 30 g/km) elasticity (minimum bending radius 2 mm); mechanical strength (stretching strength 7 y.mm^2); inexistence of mutual interference; nonconductance; blast proof; high electro insulation strength (for example, fiber of 20 cm length has? About 10000 V voltage); high corrosion resistance, especially to chemical solvents, oils, water.

In the sphere of optical communication wide bandedness and small losses are more important.

In the practice of using fiber-optical sensors the last four properties are of the greatest importance. Such properties as elasticity, small diameter and mass are very useful. Wide bandedness and small losses comparatively increase, though not always, the possibilities of optical fiber. This advantage is detected by sensor wear out. Though by widening of functional possibilities of fiber-optical sensors this situation will be slightly improved in the nearest future.

As will be shown in the following paragraph optical fiber can be used in fiber-optical sensors simply as transmission lines while can play the role of the most sensitive element of transmitter.

1.4. Fiber-Optical Sensors

Simple fiber-optical sensor, on its part, represents a light guide cut on the one end of which light source - optical diode or laser and on the second end photo detector (radiation detector) are mounted. Generally, light flux modulation takes place directly in light guide and not in light source as it happens in communication system. Any affect on optical fiber is more or less effecting light distribution in light guide. At the expense of selection of respective structures of optical fiber in sensors one type of effect is strengthened and others are weakened.

Basic principles and constructional solutions of creation of fiber-optical sensors have been for the first time formed in Georgia by Prof. J.Bakhtadze in 1965. In Japan the mentioned works were begun in 1973, and in other countries mainly after 1980.

The publications about more or less acceptable development of sensors and conformable test samples, not counting the inventions by Prof. J.Bakhtadze in 1965 and his thesis defended in 1966, have appeared in the second half of the 70th. Although it is considered that this type of sensors, as one of the trends of technique, was formed only since 1980. Thus, fiber-optical sensors are very young sphere of technique.

Consider the simplest type of fiber-optical transmitters given in Figure 1.22. It consists of: light source 1 semitransparent mirror 2, optical fiber 3 mounted on research object hoto detector 4, analog-to-digital converter 5, processor 6 with a special program processing the received signals and displaying them visually on monitor screen in graphical form. The transmitter works on light intensity variation that is conditioned by deformation of the researched object or external loads acting on it.

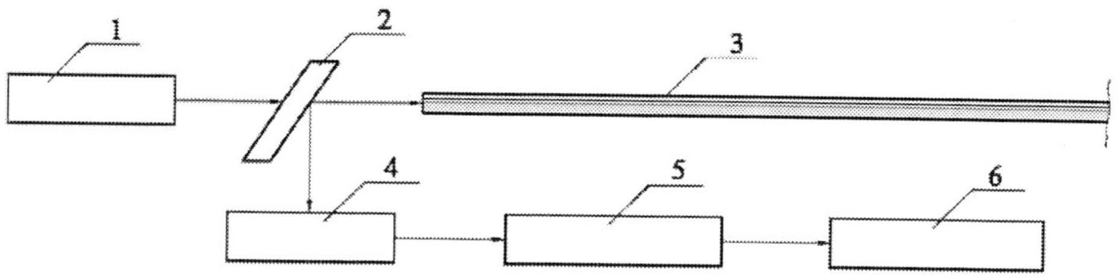

Figure 1.22. Diagram of fiber-optical transmitter.

1.5. Classification of Fiber-optical Sensors and the Examples of Their Using in Civil Engineering

The modern fiber-optical sensors measure almost everything: pressure, temperature, distance, location in space, rotation rate, linear displacement rate, acceleration, oscillation, mass, sound waves, liquid level, deformation, refractive index, electric field, magnetic field, gas concentration, radioactive radiation dose, etc.

According to optical fiber application of fiber-optical sensors are classification is in to two groups: first, of outer action where optical fiber is used only as communication lines and the second, of inner action where optical fiber performs the function of sensitive element (modulator), as well. In the latter case the relation of flux passing in optical fiber to electric field (effect), magnetic field, (effect of Faraday), vibration, temperature, pressure, deformation (for example, microbendings) is used. Some of these properties in optical communication lines are estimated as negative, while the effectiveness of their detection in transmitters is considered as positive.

According to action principles three types of fiber-optical sensors are recognized: passing (Figure 1.23), reflecting (Figure 1.24) and antenna like (Figure 1.25).

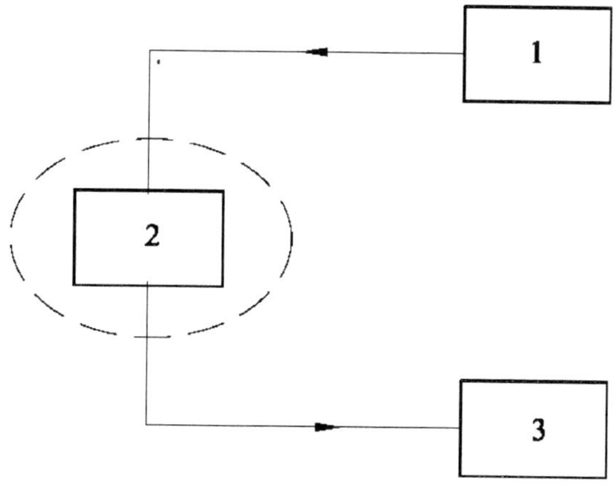

Figure 1.23. Diagram of passing type fiber-optical sensor.
1 – light source, 2 – sensitive element; 3 – detector.

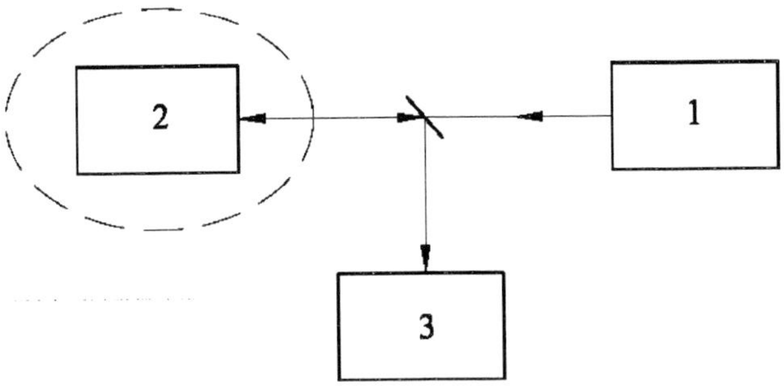

Figure 1.24. Diagram of reflecting type fiber-optical sensor.
1 – light source, 2 – sensitive element; 3 – detector.

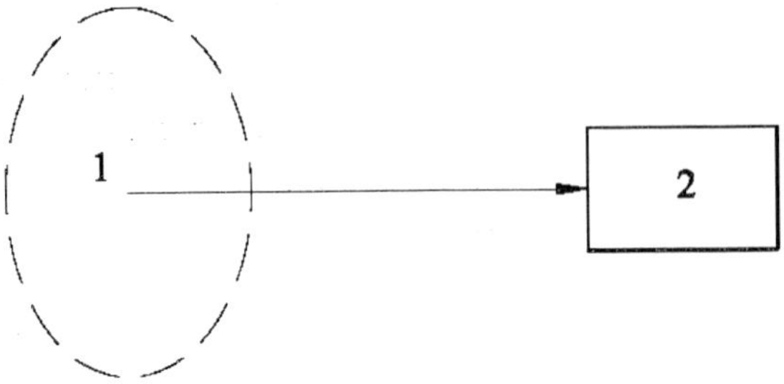

Figure 1.25. Diagram of antenna type fiber-optical sensor.
1 – light source; 2 – sensitive element.

In passing type fiber-optical sensor (Figure 1.23) the beam from light source is fed to optical fiber, the second end of which is connected to modulator. The outcoming modulated light flux on the second end with connected optical fiber is supplied to detector. In this case optical fiber is used as connection line between light source, modulator and detector. At the same time, the device is connected in series, therefore, such system is said to be of outer action and of passing type.

In the transmitter given in Figure 1.24 light propagating in optical fiber is reflected from the object and passing the same fiber light divider is supplied to detector. The presented scheme proves that fiber-optical sensor is of outer action as optical fiber is used only as communication line between object, light source and detector but it is not of passing type as in the previous case but of reflecting type as light is reflected from the object and via the same optical fiber returns back to detector.

Consider some examples of using fiber-optical sensors in civil engineering.

Diagnostic system composed of optical sensors. The research sphere is detectors of mechanical deformation and temperature variation. Composition: system contains narrow band light source which regulates the variable wave length beam and sends it to main optical fiber. Bragg grating diffraction reflecting type sensors are arranged along the whole length of optical fiber. Sensors transmit such wave lenght light beams that are corresponded with their transmission minimum. The transmitted signal changes under the effect realized on the sensor. Light source that regulates the wave length enables scanning of main (basic) light beam with its predetermined wave lengths so that to isolate wave length transmitted by each individual sensor. The light beam of the mentioned power is propagated in sensors and by way of detectors is transformed into electric signal which afterwards is processed in information processing circuit (block). The latter detects light beam power sample which is sent from detector and emits outcoming signals which transfer information about action exerted on each sensor. The system can be arranged around Fabry-Pavo circuit and at the same time provide high precision of deformation measurement. The system can be moved to different places with the aim to better supply of light into fiber and to better register the pulses reflected from transmitters.

Crack or strain monitoring. The system of crack or deformation monitoring envisaged for crack or deformation control consists of four optical fibers placed side by side and loaded with light beams. All four optical fibers are in close connection with the researched object. The fibers are placed so that to cross the line where crack is expected to originate. Light sources are placed at one of the ends of each fiber and light detectors at the other end, respectively. The variation or discontinuation of light propagated in optical fiber, caused by crack formation in construction is used for fixing the crack formation. By means of light variation or discontinuation in fibers it is possible to determine crack location or direction of its propagation.

Area controlling fiber-optical sensor. The linear sensor of the described deformation consists of jacketed optical fiber. Light passing through optical fiber is modulated in phase by deformation, the detection of which is done by interferometric reflectometer. Sensor is used for monitoring of failures, structural integrity of constructions and vehicles motion on highways.

Crack monitoring. Crack monitoring system contains semitransparent plate with channeled bottom surface. Optical fibers are placed in channels. When the plate is adjusted on the construction for monitoring, optical fibers tightly stick to its surface. By means of discontinuation of light propagation in fibers crack formation in construction is recorded. Light source can be laser or emission diode that is connected to one end of the fiber, light propagated in fiber is registered on detector connected to sensitive device. Sensitive device generates alarm signal. Alarm signal is reflected when the information exceeds the predetermined level.

Figure 1.26 shows schematic diagram of the device for monitoring of aircraft structural uniformity and Figure 1.27 shows optical block diagram. The device consists of: emitter (from light source) 1; autocollimation block 2; optical switch 3 with one end connected to autocollimation block 2 and with the other - to fiber-optical cable 4; light divider 5 optically connecting fiber-optical sensors 6 and fiber-optical cable 4; detector 7 in optical connection with autocollimation block 2, while detector 7 on its part is connected to processor 8; control block 9 and monitor 10.

The device operates as follows: fiber-optical sensors 6 are installed at prefabrication of dangerous sections (in compositional constructions) or are installed on the surface at exploitation. Emitter 1 is switched, the radiation of which in the form of pulse flow through autocollimation block 2 and optical switch 3 is introduced in the circuit of fiber-optical light guide 4 from right input. Pulse beam is distributed in the whole system.

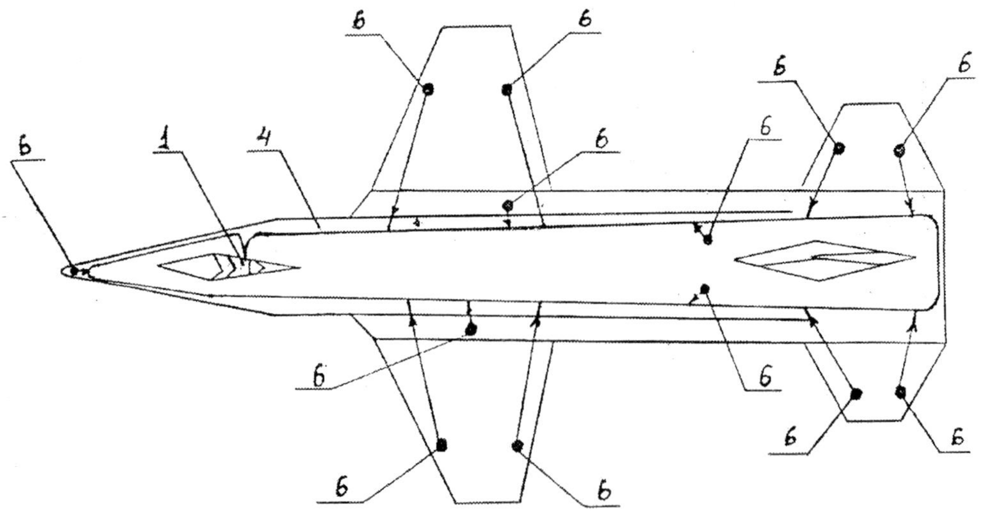

Figure 1.26. Schematic diagram of aircraft structural uniformity monitoring device.

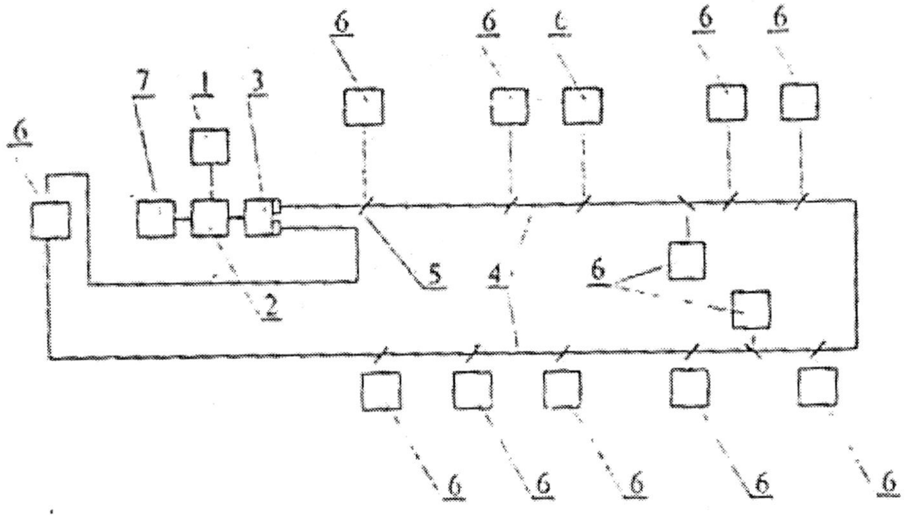

Figure 1.27. Optical block-diagram.

Signal reflected in fiber-optical transmitters 6 returns back, passes optical switch 3, autocolimation block 2 and enters detector 7, from detector 7 it is transmitted as an electric signal to processor 8 and processed signal is displayed on the monitor in the form of reflectograms. Then, by means of control block 9 optical switch 3 is switched into second left state and radiation pulse is sent to the circuit of fiber-optical light guides 3 and the monitor receives the information about temperature, oil level, pressure variation, about cracks, corrosion, wear out, All this is displayed on monitor 11 installed in pilots cabin, so that the team has the possibility to supervise the operation of the whole system during the whole period of exploitation and to transmit by Internet the information to Earth stations.

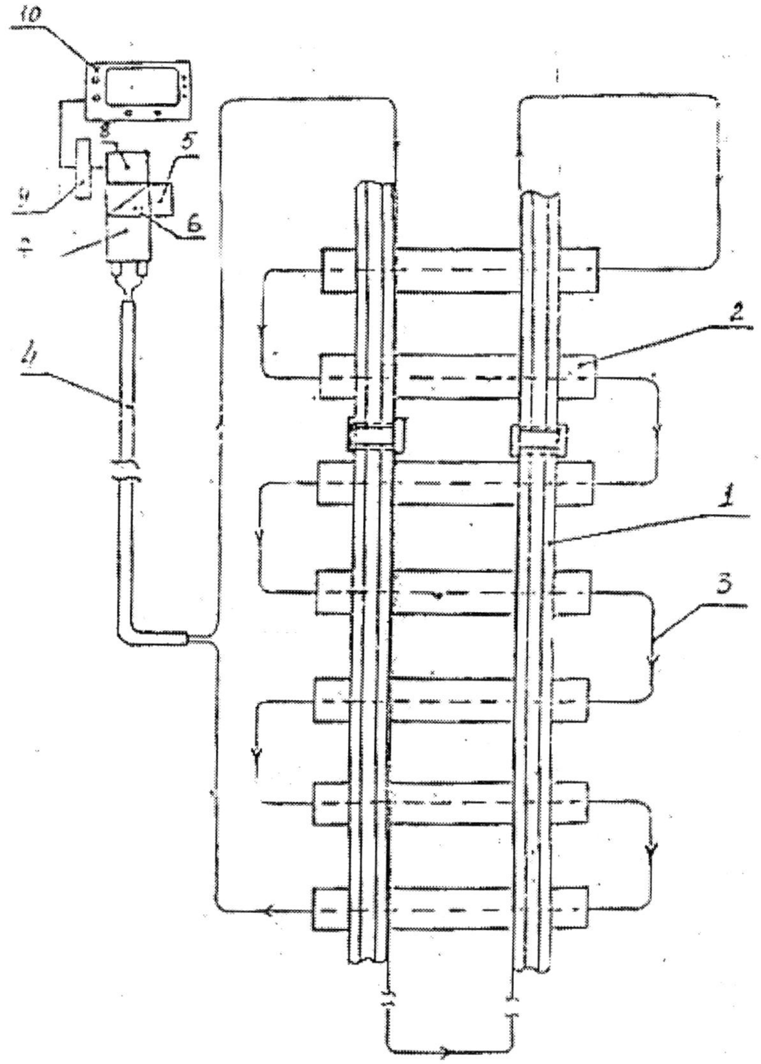

Figure 1.28. Schematic diagram of railroad track structural uniformity monitoring device.

The performance of structural integrity monitoring of aircrafts by means of the above described device guarantees the availability of timely in-flight information for the team, as well as, for ground stations about aircraft state which will decrease accidents and catastrop hes, will decrease labor capacity of control and expenses. The given device can also be used for dynamic mechanical systems where interrelating states of the bodies entering the system constantly change.

We also have elaborated the scheme of a device for railway structural integrity monitoring which is given in Figure 1.28, while Figure 1.29 shows the diagram of fiber-optical light guide arrangement on rails. The device consists of: rails 1, ties 2, fiber-optical light guides 3, fiber-optical cables 4 and light source 5.

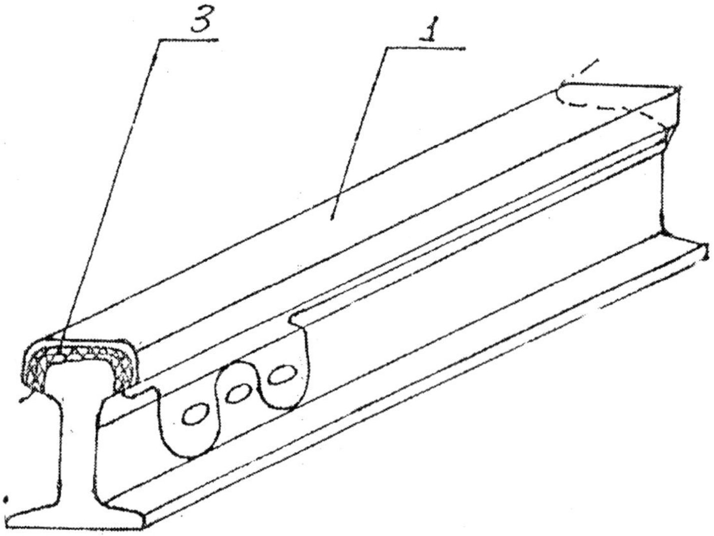

Figure 1.29. Diagram of fiber-optical light guide mounting on rails.

The device works as follows: on fabrication grooves are made on the front surface of tail ends; the diameter of grooves is a little more than the diameter of fiber-optical light guides to the configuration acceptable for rail 1 wearing. After mounting of light guide groove is filled with filler corresponding to the strength of rail material, on the border of rail 1 transmitter holes fiber- optical light guide is attached zigzag wise, also on upper shelve lower plane of rail 1 on the whole length the sensitive fiber-optical light guide 3 is attached, while light guide 3 are put in ties 2 at fabrication. All these light guides make composite fiber-optical circuit. Light source 5 is switched in, its radiation in pulse form by autocollimation block 6 and optical switch 7 are fed to the circuit of fiber-optical light guides 3 from one end of it. Beam pulse is propagated in the whole system and through the second end returns back to the second arm of optical switch 7, passes optical switch, autocollimation block 6 and gets into detector 8 where it transforms into electric signal, is transmitted to processor 9 and monitor 10 displays the respective reflectograms. Then optical switch 7 is switched into another state from where beam pulse is sent to fiber-optical light guide circuit 3 and reflectogram is received from this condition. This method enables to determine crack formation place (coordinates), direction and length. Besides, a special program is provided in the processor which as a result of the received signals gives the analysis of stress-strained state, values of wear out, geometrical dimensions and registers the formation of corrosion.

The realization of railway track control with the above described device decreases working capacity of control and expenses, enables to avoid the expected accidents, the prognosis of the railway failure becomes possible.

Because of the increase of demands in communication sphere two industrial revolutions have happened during the past 20 years: optical-electron and fiber-optical sensorization. The development of optical electronic industry enabled the creation of such product as CD players, laser printers, bar-code scanners and laser guides. Fiber-optical industry is directly connected with telecommunication renewal-development, it ensures high quality of operation, it considerably reliable and cheep.

Parallel to the mentioned direction the technology of sensors prepared of fiber became particular, as the user needed optical electronic, as well as, fiber-optical industry. The greatest part of components connected with this sphere of industry were often developed in the direction of fiber-optical sensors. On their part the development of fiber-optical sensors depended on the development and mass production of the components in order to satisfy the requirements of this sphere of industry. As the price decreased and qualitative improvement of fiber-optical sensors possibilities was realized, it became possible to change the traditional sensors measuring displacement, deformation, voltage, acceleration, pressure, temperature, humidity, oscillation, chemical composition, etc. The demand of their using increased in all spheres. The main advantages of optical fiber are: small mass, stability to electromagnetic, wide bandedness, high sensitivity, These factors were actively used for keeping the disadvantages, such as expensiveness and instability in exploitation, in shadow.

At present the situation has changed. In 1979 the price of laser diode was 3000 USD and had several hours of operation guarantee, today its price is some ten dollars and operation guarantee is 10000 hours. It is widely used in CD players, laser printers and laser guides. The price of one meter of single mode optical fiber in 1979 was 20$, today with significantly improved optical and technical characteristics it costs less than 0.1$ per meter.

Integrated optical devices, the use of which at that time was impossible, today are commonly used in fiber-optical gyroscopes. The continuation of such tendency promotes the development of fiber-optical sensors, particularly, the improvement of their physical and optical characteristics and their introduction in other spheres of human activity.

The unique properties of fiber-optical light guides are: stability to electromagnetic induction, safety to blasting, high electric insulation, high corrosion resistance, inexistence of internal interference. Particularly useful are such properties as elasticity, small diameter and mass.

The phenomena that generally negatively effect on fiber-optical communication line are often very useful for fiber-optical sensors. Fiber deflection leads to additional losses of light while randomly arising voltages - to double refraction of light that causes additional physical shift and deterioration of information channel frequency properties.

One of the disadvantages of optical fiber in communication is considered micro Leflectionson which light scattering is happening. They are the main reason of light losses in light guide. One micron length of such 100 deflections in communication line may weaken the light by some ten times. Such high sensitivity for microstrains is effectively used in fiber-optical sensors where microdeflections are preliminarily formed with the help of special devices. These devices transform the registered physical parameters into small deviations of relief plates which cause fiber deformation. Photo receiver on the light guide output registers all changes of light flow caused with purposeful microdeflections. This principle is used in order to control of stress-strained state in objects, in acoustic wave sensors for measuring magnetic, electric fields, temperature and acceleration.

As is known in hair diameter optical fiber hundreds of light pulses – modes - may propagate in the form of electromagnetic waves which provides their uniqueness compared to copper wires. As to aerial transmitters they receive or radiate light waves through open air.

Compared to the devices in use wide application of new type sensors in industry need a certain time so that their characteristics be adapted to complex situations, adaptation of customers, production of new competitive devices with respective technical characteristics, as well as, low cost.

Industrial application of fiber-optical sensors generally began since 1980. Such sensors, in the first place, gained the use in military sphere that significantly increased the popularization of fiber-optical sensors among customers. The main attention was paid to hydrophone and gyroscope researches.

The modern fiber-optical sensors can measure almost everything such as, for example, pressure, temperature, distance, position in space, deformation, displacement, voltage, rotation rate, etc.

With the view of optical fiber application if we make classification of fiber-optical sensors as was given above, they can be roughly divided into sensors where optical fiber is used as transmission line and sensors in which it is used as sensitive elements, as is given in Table 1.5. In "transmission line " sensors multi mode optical fiber is mainly used, single mode fibers are used in all other sensors.

As is seen from Table 1.5 the main elements of fiber-optical sensors are: optical fiber, light emitting and receiving devices, optical sensitive element. Besides, special lines are necessary for interconnection of these elements and formation of the sensor metering system, also computer technologies are necessary for processing and storing of received information. Thus, for practical implementation of fiber-optical sensors the systems technique elements are necessary which in interconnection with the above given elements and communication lines, make metering system.

Table 1.5.

Structure	Metering physical value	Used physical phenomenon, property	Detected value	Optical fiber	Metering parameters and peculiarities
Sensors on optical fiber in communication lines degree					
Passsing	Electric voltage, electric field	Pavelse effect	Polarization	Multi mode	1÷1000v 1÷1000v/cm
Passing	Electric current power, stress, magnetic field	Faradey effect	Polarization angle	Multi mode	±1% precision at 20-85°C
Passsing	Temperature	Variation of semiconductors absorption	Passing light intensity	Multi mode	-10÷300°C precision (±1°C)
Passsing	Temperature	Constant luminescence variation	Passing light intensity	Multi mode	0÷70°C precision (±0,04°C)
Passing	Temperature	Violation of optical path	Passing light intensity	Multi mode	"Switch in/out" mode
Passing	Hydro	Complete	Passing light	Multi mode	Sensitivity 10

	acoustic pressure	reflection	intensity		mPa
Passing	Acceleration	Photo elasticity	Passing light intensity	Multi mode	Sensitivity approximately 1
Passing	Gas concentration	Absorption	Radiated light intensity	Multi mode	Remote observation on the distance of 20 km
Reflection	Sound pressure in atmosphere	Multi component interference	Reflected light intensity	Multi mode	Sensitivity characteristic for condenser microphone
Reflection	Oxygen content in blood	Spectral changes	Reflected light intensity	Bunch like	Allowable from catheter
Reflection	Intensity of radio waves radiation	Variation of liquid crystal reflection index	Reflected light intensity	Bunch like	Aonfailure control
Antenna	Parameters of high voltage pulses	Light guide radiation	Radiated light intensity	Multi mode	front duration up to 10 nsec
Antenna	Temperature	Infra-red radiation	Radiated light intensity	Infra-red	250-1200 °C precision(±1%)
Circular interferometer	Revolution rate	effect	Light wave phase	Single mode	>0,02
Circular interferometer	Electric current power	Faraday effect	Light wave phase	Single mode	Fiber
Interferometer	Hydro acoustic pressure	Photo elasticity	Light wave phase	Single mode	1-100 rad at/m
Interferometer	Electric current power, magnetic field intensity	Magnetic penetrability	Light wave phase	Single mode	Sensitivity 10^{-9} at/m
Interferometer	Electric current power	Joule effect	Light wave phase	Single mode	Sensitivity
Interferometer	Acceleration, displacement, deformation	Mechanical compression and stretching	Light wave phase	Single mode	1000 rad/m

Table 1.5. Continued

Structure	Metering physical value	Used physical phenomenon, property	Detected value	Optical fiber	Metering parameters and peculiarities
Interferometer	Temperature, deformation	Thermal compression and stretching	Light wave phase	Single mode	High sensitivity
Interferometer	Hydro acoustic pressure	Photo elasticity	Light wave phase (interference)	Single mode	High sensitivity
interferometer	Radiation spectrum	Wave filtration	Radiated light intensity	Single mode	High sensitivity
interferometer	Pulse, blood flow rate	Doppler effect	frequency	Single mode multi mode	$10^{-4} \div 10^{8}$m/sec
Interferometer on the basis of orthogonal polarization mode	Hydro acoustic pressure, displacement, deformation	Photo elasticity	Light wave phase	Polarization preserved	Support without optical fiber
Interferometer on the basis of orthogonal polarization mode	Strss	Magnetic netrability	Light wave phase	Polarization preserved	Support without optical fiber
Noninter-ferometric	Hydro acoustic pressure, deformation	Losses at fiber micro bendings	Radiated light intensity	Multi mode	Sensitivity 100 mPa
Noninter-ferometric	Electric current power, magnetic field voltage	Faraday effect	Polarization angle	Single mode	Consideration of orthogonal modes necessary
Noninter-ferometric	Flow rate	Fiber oscillation	Intensity relation between two modes	Single mode, multi mode	>0,3 m/sec
Noninter-ferometric	Radioactive radiation dose	Light losses formation in fiber	Radiated light intensity	Multi mode	0,01-1,000
In series and parallel type	Temperature and strain propagation	Relay back scatter	Relay back scatter intensity	Multi mode	Sensitivity section 1 m

Chapter 2

Electromechanical and Fiber-Optical Sensors Joint Operation with Cables and Guys

2.1. Introduction

In order to study joint operation of electromechanical and fiber-optical sensors with cables and guys it is necessary to investigate their main schematic diagrams. Electromechanical sensors operate in harmony with cables and guys. The adjustment of fiber-optical sensors operation with cables and guys is to be investigated, new schematic diagrams are to be drawn and tested.

2.2. Joint Operation of an Electromechanical Sensor and Cable or Guy

The main elements of electromechanical sensors are a spring and contactors. The principle of their operation is described in our monograph. Electromechanical sensors function as given in the below diagrams (Figure 2.1), where 9 is a strained cable or guy, 1 and 2 are relays, 3, 3' and 3'' consist of an actuator, reducer with planetary screw transmission, spring dynamometer 5 with a cam 10, contactors 6,7 and 8 and supports 13.

Schematic diagram of automatic device is given in Figure 2.1. Here МП-100М is used as an electric motor which works on direct current Up=27 V and is denoted as 3, 1 and 2 denote relay P_1 and relay P_2 circuits.

The mentioned device has been used in composite systems of different constructions in case of static, as well as, dynamic loads.

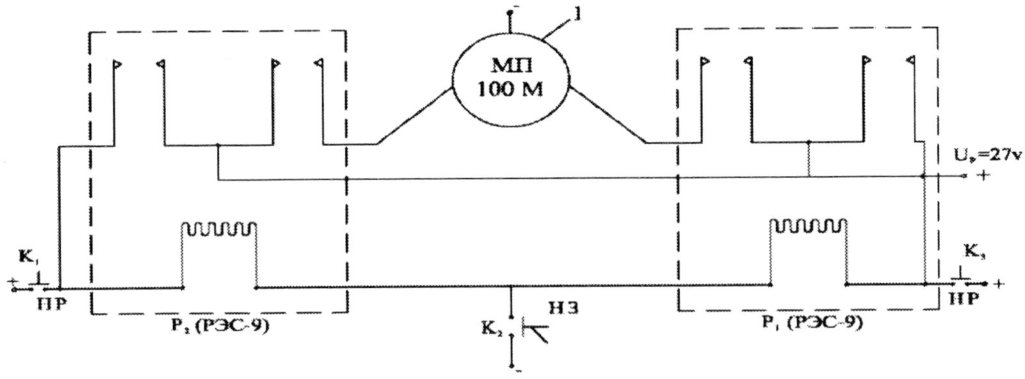

Figure 2.1. Schemalic diagram of automatic device.

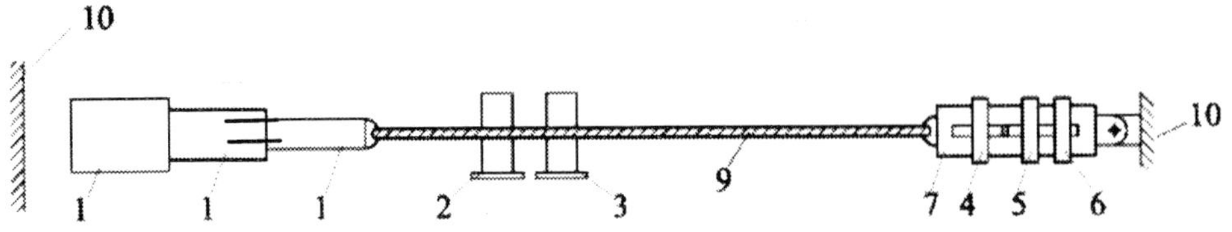

Figure 2.2. The elements of automatic strain device.

2.3. Joint Operation of a Fiber-Optical Sensor and Cable or Guy

In order to estimate joint operation of fiber-optical sensor and cable or guy it became necessary to determine their modulus of elasticity.

The determination of the modulus of elasticity for cables or rods is given in technical literature.

2.3.1. Elastic Lengthening and Modulus of Elasticity of Cables

The main exploitation advantage of steel cables is their capacity of elastic lengthening at stretching. Elastic lengthening (modulus of elasticity) of cable is characterized with relation:

$$E_k = \frac{Rl}{F\Delta l} kg / cm^2 \qquad (2.1)$$

where R is tensile stress, kg;

l is the length of the tested segment, cm;

F is cross-section area of cable, cm^2;

Δl is sample elongation at tension, cm.

The magnitude of elastic elongations of cable depends on its construction, separate technological parameters and properties of the used wire. In the average, modulus of elasticity of steel wire is accepted $E_w=2\cdot10^6$kg/cm^2.

Modulus of elasticity of cable E_c depends on the modulus of elasticity of E_w

$$E_c = a\,E_w \tag{2.2}$$

Where $a<1$.

Thus, elasticity modulus of cable is less than elasticity modulus of wire from which it is manufactured.

The more the number of repeated twisting of wire in cable and the more the angles of wire twisting to cable axis the more the decrease of cable elasticity modulus.

For determination of the modulus of elasticity of cables with point contact of wires the formulas of acad. Dinkin can be used:

for spiral cables of circular wires:

$$E_c = E_w \cos^4\varphi \tag{2.3}$$

for plain-laid cables:

$$E_c = E_w \cos^4\varphi \cdot \cos^4\varphi' \tag{2.4}$$

where φ is strand axis inclination to cable axis (single coiling)

φ′ is mean inclination of wire to strand axis.

Prof. F.V.Florinski using the data about the weight of metal part g_0 of cable at metal volume weight γ kg/cm^3, derives relation:

$$g_0=100F_c\gamma \tag{2.5}$$

where F_c is cable cross-section.

Hence

$$F_k = \frac{g_0}{100\gamma} \tag{2.6}$$

In its turn cross-section of metal part of cable proceeding from cross-sections f of wires with consideration of lays, is:

$$F_c = \frac{\sum f}{\cos\varphi\cdot\cos\varphi'} \tag{2.7}$$

or

$$\cos\varphi \cdot \cos\varphi' = \frac{\sum f}{F_k} = \frac{100\gamma \sum f}{g_0} \tag{2.8}$$

Hence:

$$E_c = \left(\frac{100\gamma f}{g_0}\right) E_w \tag{2.9}$$

It follows that the modulus of elasticity of cable made of homogeneous wires is proportional to expression $\left(\frac{\sum f}{g_0}\right)^4$.

For steel wire cable with E_w=2,2·10^{6}kg/sm^2 and γ=0,0079 kg/sm^3 we have:

$$E_c = 0{,}86 \cdot 10^6 \left(\frac{\sum f}{g_0}\right)^4 kg/sm^2$$

where$\sum f$ is given in cm^2 and g_0 - in kg/m.

Coefficients α of reduction of the modulus of elasticity E_c compared to E_w, determined by various researches in accordance to the accepted cable construction, are given in Table 2.1.

Table 2.1. The values of coefficient α determined by various researchers

Researcher	Coefficient α for cables			
	Single coiling	Double coiling	Triple coiling	Multi strand
Baumann	-	0,333	-	-
Bakh	-	0,375	-	-
Grabak	0,6	0,36	0,216	-
Dinnik	0,888	-	-	-
Dukelski	0,65-0,85	0,35-0,65*	-	0,5**
Stefan	0,65 at closed construction 0.83 at semi-closed construction			
Other researchers	0,57-0,74	0,35-0,45	0,211	-

* with organic core.

** with metal core.

Table 2.2. Modulus of elasticity of cable

Cable type (according structure)	Cable construction	E_c kg/cm$^2\cdot10^4$	
		Without prestretching	With prestretching
Spiral	1X7	148	175-210
	1X19	145	170-180
	1X37	125	153-161
	1X61	115	150
	1X75	105	145
	1X80	100	140
Plain-laid ropes with one organic core	6X7+1	85-90	102-115
	6X19+1	80-90	98-110
	8X19+1	73-75	80-85
	6X25+1	80-90	98-100
	6X37+1	70-85	85-105
	6X30+1	80-90	98-110
Plain-laid ropes with one metal core in the form of strand	6X7+(1X19)	125	140
	6X19+(1X37)	110	125
	6X37+(1X61)	98	115
Plain-laid ropes with one metal core in the form of cable	6X7+(7X7)	91	125
	6X19+(7X7)	85	110
	6X37+(7X7)	77	105
Cable-laid ropes	6X6X7	42	-

By the data of the authors researches flexible elongations of cables generally satisfy Hook's law. Generally it can be accepted that after exploitation running the cable lengthens or shortens elastically.

2.3.2. Constructional Elongations

In accordance to the accepted construction of steel cables, type and form of coiling and especially to the diameter and number of used organic or mineral cores, constructional elongations of cables after suspending vary within 0.2-4% of the length of the used cables approaching at the end of functioning to 6% and more.

Constructional elongations of cables usually happen in the initial period of exploitation which at the same time is accompanied with the increase of the modulus of elasticity of cable as was mentioned above. At initial exploitation of steel cables with organic or mineral cores there happens their final molding and in this period the cable being stretched acquires its nominal diameter, while the modulus of elasticity may increase to 20%.

In the first days of steel cables exploitation, in order to prolong their life-time, it is not advisable to bring cable load up to designed load or moreover, to overload them.

In mining industry it is recommended that after suspending the cable be several times "run" under idle load, and then make about hundred liftings at lower speed and load.

In cases when constructional elongation of cable may leed to the disturbance of standard exploitation conditions of structures where they are used (defect at bridge suspension, slackening of stretch of separate arms of towers, etc.), also in cases when it is necessary to increase the modulus of elasticity the cables are prestretched before mounting.

With this purpose, steel cables (spiral, plain-laid and cable-laid ropes with metal or organic cores) are prestretched with tensile stress not exceeding 30% of break stress of steel cables meant for using. The cables subjected to prestretching are to stay under the tension for 2-5 hours.

2.3.3. Determination of Mechanical Characteristics in Metal and Bimetal Ropes

Generally mechanical characteristics of rope are determined experimentally stretching the rope up to braking.

The given section is presenting approximated theoretical method proposed for determination of mechanical characteristics of rope based on using the diagrams of wire strain. The assumptions accepted by acad. A.N.Dinnik when deriving the approximated formula of the modulus of lateral elasticity of rope are introduced: plane cross-sections of wires stay plane even after rope strain; lateral nerrowing of wires and rope does not occur; rope core, generally consisting of one wire or, sometimes, of hemp is neglected; coil angles (wire axis deflection to rope axis) for wires of all coils of rope are accepted as similar, equal to mean angle of coil; lengths of wires of all coils of rope are considired similar; change of coil angle at rope stretching, as well as, at wire twisting is neglected.

Considering that each i-th layer of rope consists of wires of equal diameter and of the same material, in the general case differing from wire material of other layers, on the basis of the accepted assumptions the design diagram of the rope of single coiling of length ℓ is to be presented as symmetrical internally statically undefinable hinge system ABC (Figure 2.3).

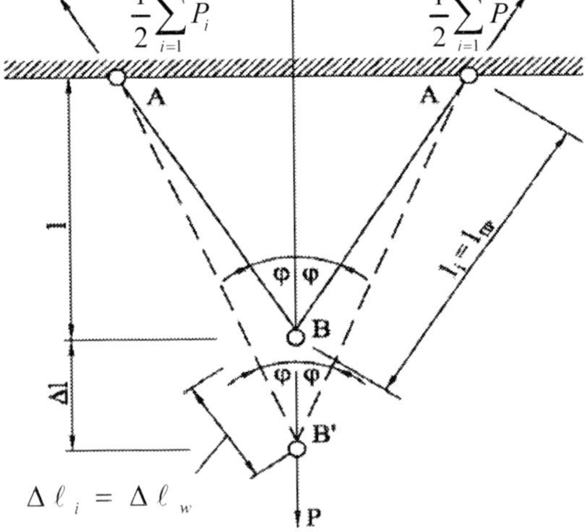

Figure 2.3. Desing diagram of a rope.

For this system the equilibrium equation is written as:

$$P-\sum_{i=1}^{i=n} P_i \cos\varphi = 0 \tag{2.10}$$

where P is load capacity of a rope;

P_i is load falling on all wires of i-th layer acting along wire axis;

$\sum_{i=1}^{i=n} P_i$ is load falling on wires of all layers of rope from 1 to n, acting along wire axis;

φ is mean angle of coils determined as arithmetic mean of all deflection angles of wires of all layers of rope.

As the lengths ℓ_w of wires of all layers of rope are considered as equal, i.e. $l_1=l_2=...=l_n=l_i=l_w$, then from design diagram of rope follow the equalities of absolute elongations Δl_w of wires of all coils of the rope, and also the equality of relative elongations ε_w of wires of all coils of the rope.

$$\Delta l_n = \Delta l_i = \Delta l_w; \quad \varepsilon_n = \varepsilon_i = \varepsilon_w \tag{2.11}$$

The equation of the consistency of displacements at absolute elongation of rope Δl, i.e. when system ABC goes over into condition ABC, is written as:

$$\Delta l = \frac{\Delta l_w}{\cos\varphi} \tag{2.12}$$

Dividing in equation 2.12 the left part on l and the right part on $l=l_w\cos\varphi$, we receive the formula estimating the connection between relative lengthening of rope ε and relative lengthening of wires:

$$\varepsilon = \frac{\Delta l}{l} = \frac{\varepsilon_w}{\cos^2\varphi} \tag{2.13}$$

In order to derive the formulas of rope load capacity and tensions $\sigma=\sigma(\varepsilon)$ in rope combine (Figure 2.4) tension diagrams similar scales of $\sigma_i=\sigma_i(\varepsilon_w)$ of wires composing each i-th coil of rope (in Figure 2.4 the number of rope coils is taken as n=3). The possibility of combination of wire tension diagrams is supported with equality $\varepsilon_i=\varepsilon_w$.

The layer, the wires of which have the least relative elongations ε_{pri} at proportional limit σ_{pri} (ε_{pri}) is called layer k and the layer, the wires of which have the least relative elongation ε_{ti} at ultimate strength $\sigma_{ti}(\varepsilon_{ti})$ - is called layer c (as is shown in Figure 2.3. $\varepsilon_{prk}=\varepsilon_{pr2}$ and $\varepsilon_{tc}=\varepsilon_{tc3}$).

It is clearly evident that any i-th layer from 1 to n, can be layer k or layr c. In a particular case, layer k can, at the same time, be as layer c.

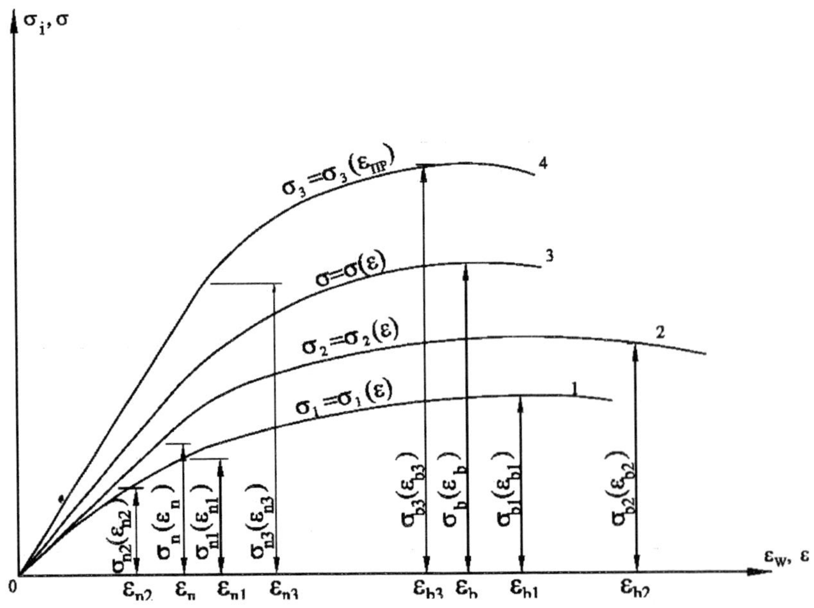

Figure 2.4. Composite diagrams of wire tenstions and theoretical diagram of rope tensions.
1 - wires of coil 1; 2 – wires of coil 2; 3 - rope; 4 – wires of coil 3.

Assuming that load P_i falling on all wires of i-th coil of rope can be expressed through tensions $\sigma_i(\varepsilon_w)$ in the wires of this coil and area F_i of cross-section of all wires of this coil by formula

$$P_i = \sigma_i(\varepsilon_w)F_i \tag{2.14}$$

from equation (1) we receive the formula for determination of rope load capacity:

$$P = \sum_{i=1}^{i=n} \sigma_i(\varepsilon_w)F_i \cos\varphi \tag{2.15}$$

In composite diagrams of wire tensions setting the different values of $\varepsilon_w < \varepsilon_{tc}$ we find from these diagrams tensions $\sigma_i = \sigma_i(\varepsilon_{pri})$ in wires of every coil of rope. Substituting values $\sigma_i(\varepsilon_{pri})$ into formula (2.15) we receive load P acting on rope.

Substituting into formula (2.15) tensions $\sigma_i = \sigma_i(\varepsilon_{tc})$ found from diagrams of wire tensions at $\varepsilon_{pri} = \varepsilon_{tc}$ we obtain breaking load P_{tc} for rope.

Formula 2.15 allows to draw diagram of rope elongation, i.e. diagram of its load capacity. Absolute lengthening of rope corresponding to load P is determined by formula:

$$\Delta l = \frac{l}{\cos^2\varphi}\varepsilon_{pri} \tag{2.16}$$

resulting from formula (2.13).

Dividing the left and right parts of formula 2.15 by rope cross-section area we get:

$$F = \sum_{i=1}^{i=n} F_i \cos\varphi \tag{2.17}$$

which because of wire incline is more than area $\sum_{i=1}^{i=n} F_i$ of cross-section of wires of all coils of rope, we receive the formula for stress determination in rope:

$$\sigma = \frac{\sum_{i=1}^{i=n} \sigma_i \left(\varepsilon_{pr}\right) F_i}{\sum_{i=1}^{i=n} F_i} \cos^2\varphi \tag{2.18}$$

Formula 2.18 allows to draw the diagram of rope stresses. As in formula 2.15 in formula 2.18 stresses $\sigma_i(\varepsilon_{pri})$ should be determined from composite diagrams of wire stresses. Relative elongation of rope corresponding to stress $\sigma=\sigma(\varepsilon)$ is determined by formula 2.13.

If at drawing diagram $\sigma=\sigma(\varepsilon)$ the scales for σ and ε are taken the same as in composite diagrams of wire stresses, theoretical diagram $\sigma=\sigma(\varepsilon)$ can be drawn on composite diagrams of wire stresses (see Figure 2.3).

As relative elongations of rope wires are similar then, as it follows from formulas 2.8 and 2.13 proportionality limit of rope $\sigma_{pr}=\sigma_{pr}(\varepsilon_{pr})$ and relative elongation of rope ε_{pr} at proportionality limit are to be calculated by formulas:

$$\sigma_w = \frac{\sum_{i=1}^{i=n} \sigma_i \left(\varepsilon_w\right) F_i}{\sum_{i=1}^{i=n} F_i} \cos^2\varphi \tag{2.19}$$

$$\varepsilon_w = \frac{\varepsilon_w}{\cos^2\varphi} \tag{2.20}$$

where $\sigma_i(\varepsilon_{wk})$ are streses in the wires of i-th coil of rope, corresponding to relative elongation $\varepsilon_{pr}=\varepsilon_{wk}$. Ultimate strength $\sigma_t=\sigma_t(\varepsilon_t)$ of rope and its relative elongation ε_t at breaking point are determined by formulas:

$$\sigma_t = \frac{\sum_{i=1}^{i=n} \sigma_i \left(\varepsilon_{tc}\right) F_i}{\sum_{i=1}^{i=n} F_i} \cos^2\varphi \tag{2.21}$$

$$\varepsilon_t = \frac{\varepsilon_{tc}}{\cos^2 \varphi} \tag{2.22}$$

respectively following, as formulas 2.19 and 2.20, from formulas 2.18 and 2.13. stresses $\sigma_i(\varepsilon_{tc})$ in wires of i-th coil of rope correspond to respective elongation $\varepsilon_{pr} = \varepsilon_{tc}$.

On the basis of Hook law the following dependences have place:

$$\sigma_i(\varepsilon_{wk}) = \mathrm{E}_i \varepsilon_{wk}, \quad \sigma_w = \mathrm{E}\varepsilon_w \tag{2.23}$$

where E_i is the modulus of longitudinal elasticity of i-th coil wires; E is the modulus of longitudinal elasticity of rope.

Therefore, from formila (2.19), considering formula (2.20) and relations (2.23) we come to approximated formula of Acad. A.N.Dinnik:

$$E = \frac{\sum_{i=1}^{i=n} E_i F_i}{\sum_{i=1}^{i=n} F_i} \cos^4 \varphi \tag{2.24}$$

All the received formulas can be applied to the constructions of ropes with single (spiral) winding.

If in design diagram of rope (see Figure 2.5) denote through ψ incline angle to strand axis, and through φ mean angle of wind of strand, i.e. the angle of incline of axes of strand wires to strand axis, then in analogy with the formulas for single winding rope one can receive formulas for double winding rope.

The main formulas for double winding rope, the construction of which, as usual, consists of similar wire of one and the same material:

$$P = \sigma_i(\varepsilon_{pr}) \sum_{i=1}^{i=n} F_i \cos\varphi \cdot \cos\psi, \quad \Delta l = \frac{l}{\cos^2 \varphi \cos^2 \psi} \varepsilon_{pr} \tag{2.25}$$

$$\sigma = \sigma_i(\varepsilon_{pr}) \cos^2 \varphi \cdot \cos^2 \psi, \quad \varepsilon = \frac{\varepsilon_{pr}}{\cos^2 \varphi \cdot \cos^2 \psi} \tag{2.26}$$

$$E = E_i \cos^4 \varphi \cdot \cos^4 \psi \tag{2.27}$$

Similarly, the approximated formulas for triple winding ropes can be derived.

Examples

Construct the diagrams of stresses $\sigma = \sigma(\varepsilon)$ and determine breaking loads P_t for metal and bimetal ropes of high voltage transmission line.

Metal Rope. The material of all wires is a special alloy; rope coil number - 3; number of wires together with core 1+6+12+18=37; diameter of each wire is d=3,85mm; wire cross-section area of wire is f=11,65mm^2. Cross-section area of all wires excluding core wire is

$$\sum_{i=1}^{i=3} F_i = 11,65 \cdot 36 = 419 mm^2$$. Mean angle of winding is φ=14°.

Bimetal Rope. The material of core is steel; coil number of core – 2; number of wires together with core wire 1+6+12=19; diameter of core wire d_0=3,1mm; diameter of each wire of core d_{st}=2,7mm; cross-section area of each core wire f_{st}=5,73mm^2; cross-section area of all wires of core $\sum_{i=1}^{i=2} F_{ist} = 5,73 \cdot 18 = 103 mm^2$. Material of shell wires – aluminium; number of shell coils – 3; number of shell wires 19+27+32=78; diameter of each wire of the shell d_{al}=2,36mm; cross-section area of each wire of shell $f_{al} = 4,37 mm^2$; cross-section area of all wires of shell $\sum_{i=1}^{i=3} F_{ist} = 4,37 \cdot 78 = 341 mm^2$. Cross-section area of wires of all coils of rope, excluding core wires, $\sum_{i=1}^{i=5} F_i = 103 + 341 = 444 mm^2$; mean angle of winding φ=14°.

Stress diagrams of rope wires of these constructions received in mechanical laboratory of the Leningrad Institute of Aircraft Instrument-Making are given in Figure 2.5 and 2.6. The Figures also give theoretical diagrams of rope stresses, the data for the diagrams being calculated by the given formulas.

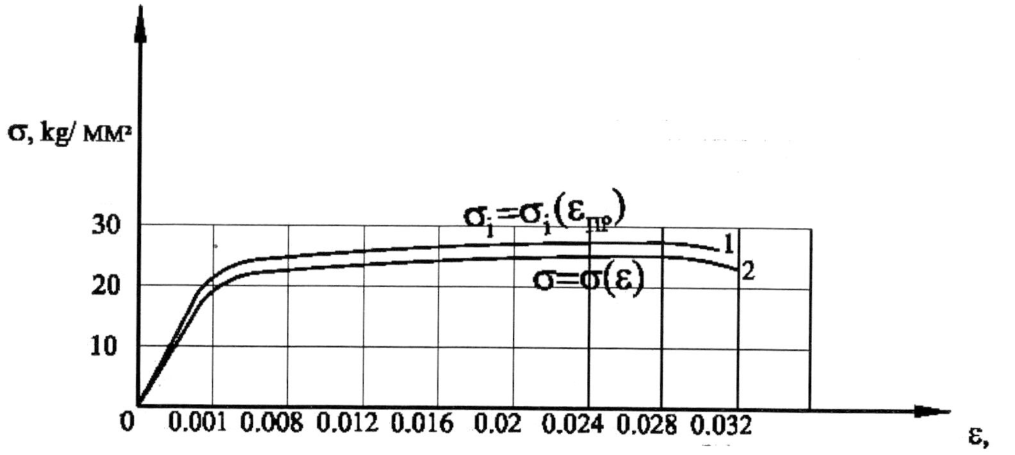

Figure 2.5. Metal rope. 1 – wire; 2 – rope.

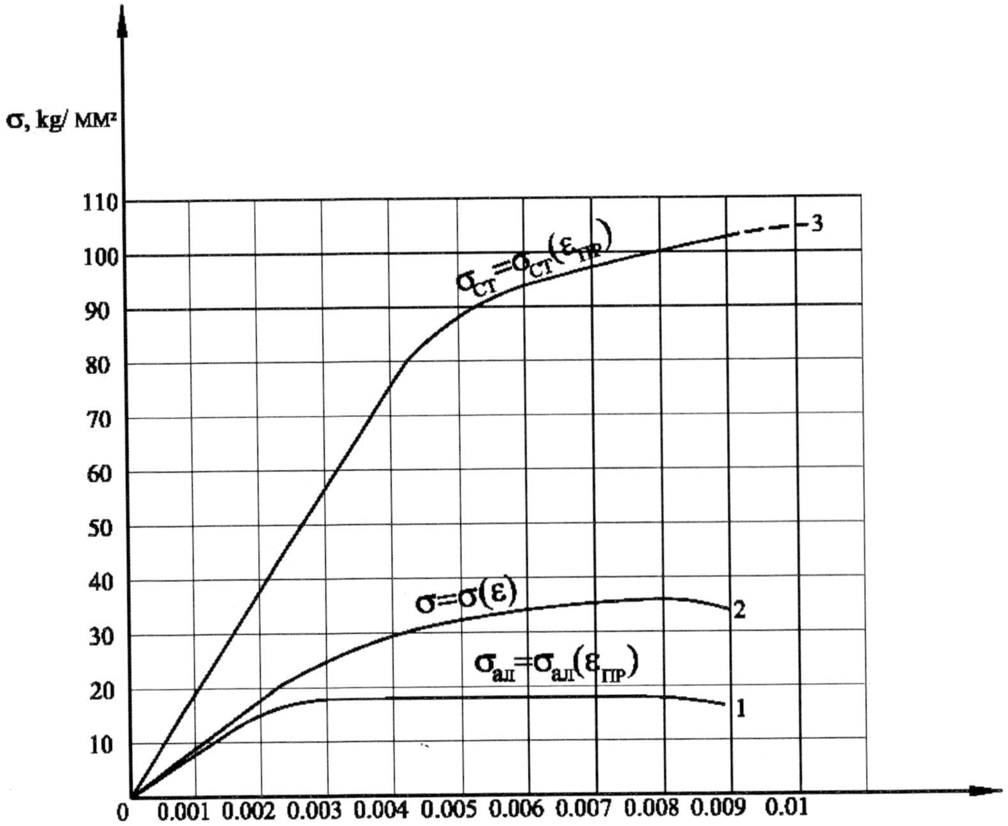

Figure 2.6. Bimetal rope. 1 – aluminium wires; 2 – rope; 3 – steel wires.

The mechanical characteristics of rope wires determined from the diagrams of wire stresses are given in Table 2.3. and 2.4. The results of mechanical characteristics of ropes, received theoretically and experimentally, are compared. The tests with the ropes of these constructions were carried out in Mechanical laboratory of the Leningrad Institute of Railway Transport Engineers.

Table 2.3.

Material	Characteristics					
	ε_{wi}	$\varepsilon_{wi}(\varepsilon_{wi})$ kg/mm^2	E_i, kg/mm^2	ε_{ti}	$\sigma_{ti}(\varepsilon_{ti})$ kg/mm^2	P_{ti} kg
Metal rope wires						
Special alloy	0,00283	18,5	6540	0,0272	27,0	314
Bimetal rope wires						
Steel	0,004	76,0	19000	0,0814	122,2	700
Aluminium	0,00187	13,1	7000	0,0080	18,3	80

Table 2.4.

Results	Characteristics					
	ε_w	$\varepsilon_w(\varepsilon_w)$ kg/mm^2	E, kg/mm^2	ε_t	$\sigma_t(\varepsilon_t)$ kg/mm^2	P_t kg
Metal rope						
Theory	0,00301	17,4	5790	0,02891	25,4	10974
Test	0,00235	13,5	5745	0,0256	26,7	11520
Difference, %	22	22	1	12	5	5
Bimetal rope						
Theory	0,001987	17,2	8662	0,008502	35,5	16245
Test	0,00185	14,0	7568	0,00785	36,7	16800
Difference, %	7	19	13	8	4	4

Let's give calculations for bimetal rope. From the diagrams of stresses of wires (Figure 2.6) it follows that rectilinear section of the diagram of rope stress will be violated in the case if in aluminium wires proportionality limit will set in $\sigma_{w,al}=\sigma_{w,al}(\varepsilon_{w,al})=13,1$kg/mm^2 and relative elongations of steel and aluminium wires will be equal to $\varepsilon_{pr}=\sigma_{w,al}=\sigma_{wk}=0,00187$, then by formulas (2.19) and (2.20) we shall respectively receive the data given in Table 2.3.

$$\sigma_w = \frac{\sigma_{st}\left(\varepsilon_{w,al}\right)\sum\limits_{i=1}^{i=2} F_{ist} + \sigma_{al}\left(\varepsilon_{w,al}\right)\sum\limits_{i=1}^{i=3} F_i}{\sum\limits_{i=1}^{i=5} F_i}\cos^2\varphi = \frac{35,53\cdot 103 + 13,1\cdot 341}{444}0,970^2 = 17,2kg/mm^2$$

having in the view

$$\sigma_{st}\left(\varepsilon_{w,al}\right) = E_{st}\varepsilon_{w,al} = 19000\cdot 0,00187 = 35,53kg/mm^2$$

$$\varepsilon_w = \frac{\varepsilon_{w,al}}{\cos^2\varphi} = \frac{0,00187}{0,970^2} = 0,001987$$

By formula (2.24)

$$E = \frac{E\left(\varepsilon\right)\sum\limits_{i=1}^{i=2} F_i + E\left(\varepsilon\right)\sum\limits_{i=1}^{i=3} F_i}{\sum\limits_{i=1}^{i=5} F_i}\cos^4\varphi = \frac{1900\cdot 103 + 7000\cdot 341}{444}0,970^4 = 8662kg/mm^2$$

Ultimate strength of rope will set in in the case when stresses in aluminium wires achieve $\sigma_{t,al}=\sigma_{t,al}(\varepsilon_{t,al})=18,3$kg/mm^2 and relative elongations of steel and aluminium wires will be equal to $\varepsilon_{pr}=\varepsilon_{t,al}=\varepsilon_{tc}=0,0080$. Then, as it follows from diagrams of wire stresses, stresses in steel wires wll be $\sigma_{st}=\sigma_{st}(\varepsilon_{t,al})=102$kg/mm^2. Hence, by formulas 2.19 and 2.20 we get:

$$\sigma_t = \frac{\sigma_{st}\left(\varepsilon_{t,al}\right)\sum_{i=1}^{i=2} F_{ist} + \sigma_{al}\left(\varepsilon_{t,al}\right)\sum_{i=1}^{i=3} F_i al}{\sum_{i=1}^{i=5} F_i}\cos^2\varphi = \frac{102\cdot 103 + 18,3\cdot 341}{444}0,970^2 = 35,5kg/mm^2$$

$$\varepsilon_t = \frac{\varepsilon_{t,al}}{\cos^2\varphi} = \frac{0,0080}{0,970^2} = 0,008502$$

Intermediate values of stresses in rope and also relative elongations of rope when these tensions arise are calculated by formulas (2.18) and (2.13). Thus, for example, when $\varepsilon_{st}=\varepsilon_{al}=\varepsilon_{pr}=0{,}0060$ from wire stress diagrams we have:

Then

$$\sigma_{st}(\varepsilon_{pr}) = 93kg/mm^2 \text{ and } \sigma_{al}(\varepsilon_{pr}) = 18,2kg/mm^2$$

$$\sigma = \frac{\sigma_{st}\left(\varepsilon_{pr}\right)\sum_{i=1}^{i=2} F_i + \sigma_{al}\left(\varepsilon_{pr}\right)\sum_{i=1}^{i=3} F_i}{\sum_{i=1}^{i=5} F_i}\cos^2\varphi = \frac{93\cdot 103 + 18,2\cdot 341}{444}0,970^2 = 33,4kg/mm^2$$

$$\varepsilon = \frac{\varepsilon}{\cos^2\varphi} = \frac{0,0060}{0,970^2} = 0,006376\,.$$

Load P_t causing rope break is found using Formula 2.15

$$P_t = \left[\sigma_{st}\left(\varepsilon_{t,al}\right)\sum_{i=1}^{i=2} F_{ist} + \sigma_{al}\left(\varepsilon_{t,al}\right)\sum_{i=1}^{i=3} F_{ial}\right]\cos\varphi = \left(102\cdot 103 + 18,3\cdot 341\right)\cdot 0,970 = 16245kg$$

2.3.4. Modulus of Elasticity of the Inclined Cable

Determination of the Modulus of Elasticity

The modulus of elasticity is determined by using a gauge length of not less than 100 in (2.54 m) and a gross metallic area, including zinc coating of the strand conforming to ASTM A-586 Specifications. The elongation readings used for computing the modulus of elasticity are taken when the strand or rope is stressed to not less than 10 % of the minimum rated breaking strength, or more than 90 % of the prestretching load. The modulus of elasticity shall not be less than 24 x 10^6 psi (16.9 x 10^9 kg/m^2) for $\frac{1}{2}$ - $2\frac{9}{16}$ in (12.7 – 65.1 mm) nominal diameter strand and 23X10^6psi (1.62X10^9kg/m^2) for $2\frac{5}{8}$ in (66.6 mm) and larger nominal diameter strand. These values are for normal prestretched helical type strands. For a parallel-wire strand the modulus of elasticity is in the range of $28X10^6$ – $28.5X10^6$psi ($19.7X10^9$ – $20X10^9$kg/m^2)

The modulus of elasticity E of the rope is low for low loads and increases as the load is increased into the normal working range. Creep may occur for sustained loads.

For Julicher Bridge in Dusseldorf long-term elongation was reported at 0.25 x the ultimate equivalent to E=$16.9X10^2$ ksi[6]. For short term loads from 0.25 to 0.40 x the ultimate value of E was $23.1X10^3$ ksi. These figures appear to be typical of German practice. The design of the Usk River Bridge was based on a value of E equal to $22.8X10^3$ ksi.

The Apparent Modulus of Elasticity

The stiffness of the cable-stayed bridge depends largely upon the tensile stiffness of stay cables.

The displacement of the end of the free hanging cable under an axial load depends not only on the cross-section area and the modulus of elasticity of the cable but to a certain extent on the cable sag, as proved by Ernst.

Let us consider an inclined cable connected by a hinge at the lower support and with one movable bearing as the upper support (Figure 2.7). Due to the action of force F the cable takes the shape of a chain line, where L>l and L is the length of chain line between supports. If F became infinite the cable would be straight. The end C moves to the position C_1 and the expansion is

$$\Delta l = L - l \tag{2.28}$$

By increasing the load to

$$F_1 = F + \Delta F \tag{2.29}$$

the end C moves the distance

$$\Delta\Delta l = \Delta l - \Delta l_1 \tag{2.30}$$

and elongation is

$$\varepsilon_f = \Delta\Delta l / l \tag{2.31}$$

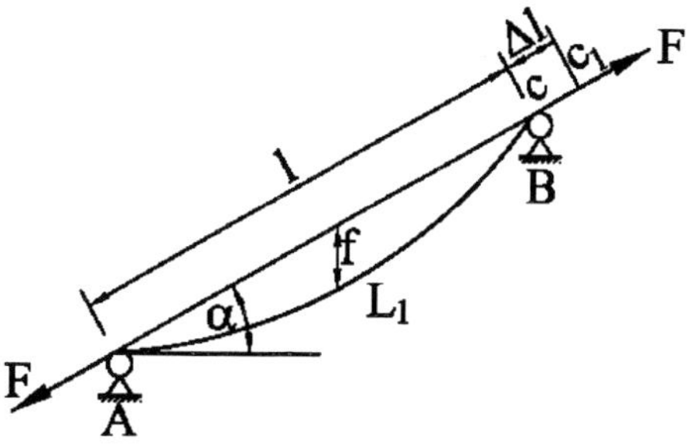

Figure 2.7. The inclined cable arrangement.

Therefore, the apparent modulus of elasticity may be expressed as:

$$E_f = \sigma / \varepsilon_f \tag{2.32}$$

The ideal modulus of elasticity

The cable under an axial load undergoes an elastic stretch:

$$\varepsilon_e = \sigma / E_e \tag{2.33}$$

where E_e indicates the corresponding modulus of elasticity.

An ideal or equivalent modulus of elasticity of cable E_i is defined as depending on moduli E_f due to sag and E_e due to elasticity, or

$$E_i = \frac{\sigma}{\varepsilon_f + \varepsilon_e} \tag{2.34}$$

where

$$\varepsilon_f = \frac{\sigma}{E_f} \quad \text{and} \quad \varepsilon_e = \frac{\sigma}{E_e} \tag{2.35}$$

From the above, we obtain:

$$E_i = \frac{E_f E_e}{E_f + E_e} \tag{2.36}$$

To find E_i, Δl should be determined. The comparison between the catenary and parabola indicates negligible difference. Therefore, the catenary may be satisfactorily approximated over this length by a parabola (Figure 2.8)

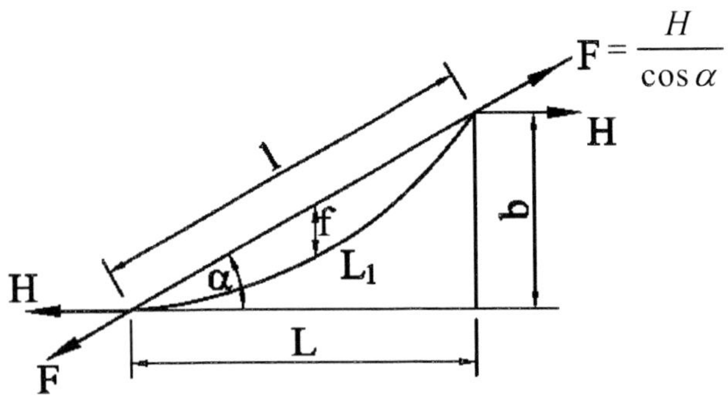

Figure 2.8. Parabolic cable.

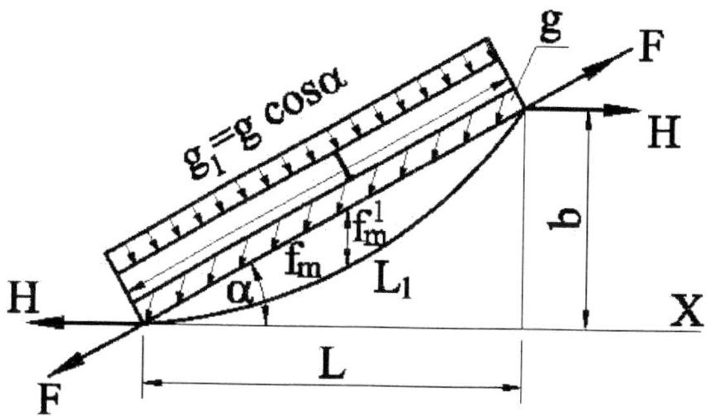

Figure 2.9. Uniformly loaded cable.

Let us now consider an inclined cable under uniformly distributed load (Figure 2.9). At the location $x=x_m/l$:

$$M_s=Hf_m \qquad (2.37)$$

where H is the horizontal component of cable tension, f_m is the ordinate to the cable curve measured downward from the chord, and M_s is the simple beam moment for the given span and load.

Under load g_1

$$M_g=g_1l^2/8$$

Because $g_1=g\cos\alpha$ and

$$l = L/\cos\alpha \qquad M_g = gL^2/(8\cos\alpha)$$

Condition $M_s=M_g$ yields

$$Hf_m = gL^2/(8\cos\alpha)$$

and if H/g=h, we have

$$f_m = L^2/(8h\cos\alpha) \qquad (2.38)$$

The cable length is

$$L_1 = l + \frac{8}{3}\frac{\left(f_m^{'}\right)^2}{l}$$

By neglecting the high order terms, we obtain:

$$\Delta l = L_1 - l = \frac{8}{3}\frac{\left(f_m^{'}\right)^2}{l}$$

By designating

$$f_m^{'} = f_m \cos\alpha = L^2 / 8h \text{ and } \quad l\text{=L/cos}\alpha$$

we obtain

$$\Delta l = \frac{L^3 \cos\alpha}{24h^2} \tag{2.39}$$

By substituting into the eq. (2.45) the values
L=*l*cosα h=H/g H=Fcosα h= Fcosα/g
we obtain

$$\Delta l = \frac{g^2 l^3 \cos^2\alpha}{24F^2} = CF^{-2} \tag{2.40}$$

and

$$\frac{d\Delta l}{dF} = -2CF^{-3} = \frac{g^2 l^3 \cos^2\alpha}{12F^3} \tag{2.41}$$

Generally

$$E = \frac{\sigma}{\varepsilon} = \frac{Fl}{A\Delta l} = \frac{l}{A}\frac{dF}{d\Delta l}$$

By substituting for dF/dΔ*l* its value from (2.41) we obtain:

$$E_f = \frac{12lF^3}{Ag^2 l^3 \cos^2\alpha} = \frac{12F^3}{Ag^2 L^2}$$

By designating

g/A=γ

we have

$$E_f = \frac{12F^3}{Ag^2L^2} = \frac{12\sigma^3}{(\gamma L)^2} \tag{2.42}$$

By introducing (2.42) into (2.36) we obtain the expression for the ideal modulus E_i:

$$E_i = \frac{E_e}{1+\left[(\gamma L)^2 / 12\sigma^3\right]E_e} \tag{2.43}$$

where

E_i is Young's modulus of cable having sag;
E_e is Young's modulus of straight cable;
γ is specific weight of the cable;
L is horizontal length of the cable;
σ is tensile stress in the cable.

We assume for straight locked-coil steel wire rope Young's modulus equal to E_e=10800 tons in^2 and the approximate value:

$$\gamma = \frac{g}{A} = 1.37\times10^{-4} tons / in^3$$

Then

$$\gamma^2 E_e = 1.37^2 \times 10^{-8} \times 10800 = 2.02\times10^{-4} tons / in^2$$

$$\frac{(\gamma L)^2}{12} E_e = \frac{2.02\times10^{-4}\times L^2 \times 12^2}{12} = 2.42\times10^{-3} L^2 tons / in^2$$

Therefore,

$$E_i = \frac{10800}{1+(2.42\times10^{-3}\times L^2)/\sigma^3} tons / in^2 \tag{2.44}$$

where L is in ft and σ is in tons/in^2.

Assuming numerical values for span L and stresses σ its is possible to determine from formula (2.44) different values of the ideal modulus E_i shown in the diagram (Figure 2.10).

Static calculations for the live load are based on the idealized modulus of elasticity E_i which decreases as the length of cable increases. If the load on the sloping cable is increased, its sag is reduced and its ends move away from each other. Solely from this elongation of the chord an apparent Young's modulus can be derived which increases with load increase. This effect, together with the elastic deformation of cable, can be used for calculation of an idealized modulus of elasticity which is then introduced into static calculations.

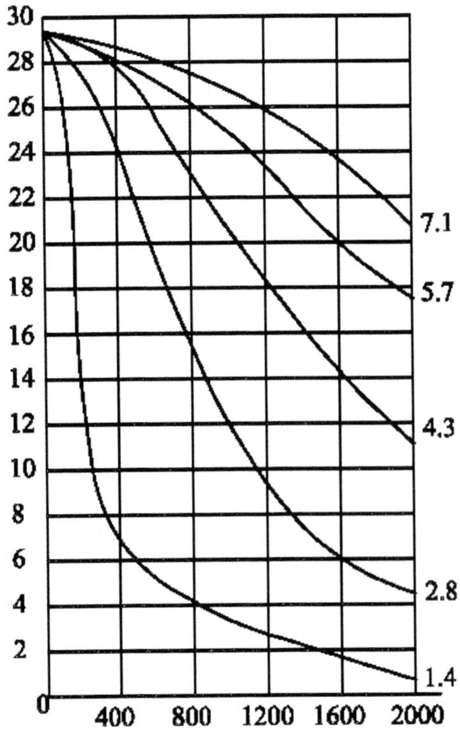

Figure 2.10. Variation of ideal modulus with span length.

In Figure 2.8 this modulus is diagrammatically shown on the ordinate as a function of the cable stress and the horizontal distance between the tower and the anchor of the stay cable is shown on the abscissa. For very long bridges the loss of E_i can be as large as 40 %.

The economical limit for cable lengths for inclined cable systems is therefore between 658 and 987 ft (200 m and 301 m). Nonetheless, longer lengths of cables could be subdivided by intermediate supports to avoid this disadvantageous effect, but it is debatable how far such a design could be made to look attractive.

It is certain, however, that even with longer cables, the inclined cable bridge could still successfully compete with the conventional suspension bridge.

2.3. Determination of the Modulus of Elasticity of Small Diameter Optical Fibers and Cables

Test Results of the Modulus of Elasticity of Optical Fibers and Cables

In order to estimate the modulus of elasticity of optical fibers, high and low modulus quartz fibers of two diameters 1.3 mm and 2.3 mm were used.

High modulus fiber with diameter 1.3 mm gave mean elasticity modulus E_{fib}=751880kg/cm^2 that practically is very close to the data by L.S.Grattan and B.T.Meggitt - E=730000kg/cm^2.

Low modulus optical fiber rod Φ=1,3mm was tested several times and mean elasticity modulus in one case was E=108916kg/cm^2 and in the second case - E=138742,41kg/cm^2.

Here it should be mentioned that elasticity modulus for optical fiber rod, as seen from diagram (Figure 2.11) changes with the increase of load and for calculations its averaged values are used.

In order to estimate elasticity moduli of cables different material and diameter cables were tested, particularly: 2.08 mm, 2.1 mm, 2.5 mm 3.5 mm and 4.5 mm.

Some cables were tested on stretching machine and some – on a specially constructed test bench (see Figure 2.12, 2.13). In both cases, for optical bar as well as for cables, tensometers MK-3 were used. The procedure of multiple loading-unloiading was carried out.

Tensometers were mounted in several places along the test member. As in the case of optical fiber bar, in the case of cables as well, elasticity modulus changes with strain and changes also along the test member.

Joint Operation of Optical Fiber Bar and Cable

For estimation of joint operation of optical fiber bar and cable, the low modulus optical fiber (of quartz material) ∅ 2.3 mm and the zinc coated cable ∅ 22.08 mm have been chosen connected to each other with clamps, as well as with sticky tapes. The middle part of optical fiber rod was of sinusoidal form with length -100 mm and arm -10 mm.

Quart optical fiber ϕ2,3mm

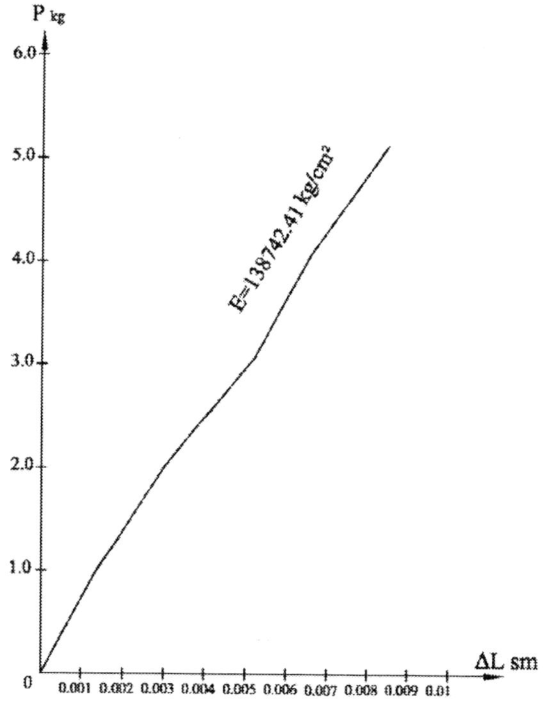

Figure 2.11. Quart optical fiber ϕ2,3mm.

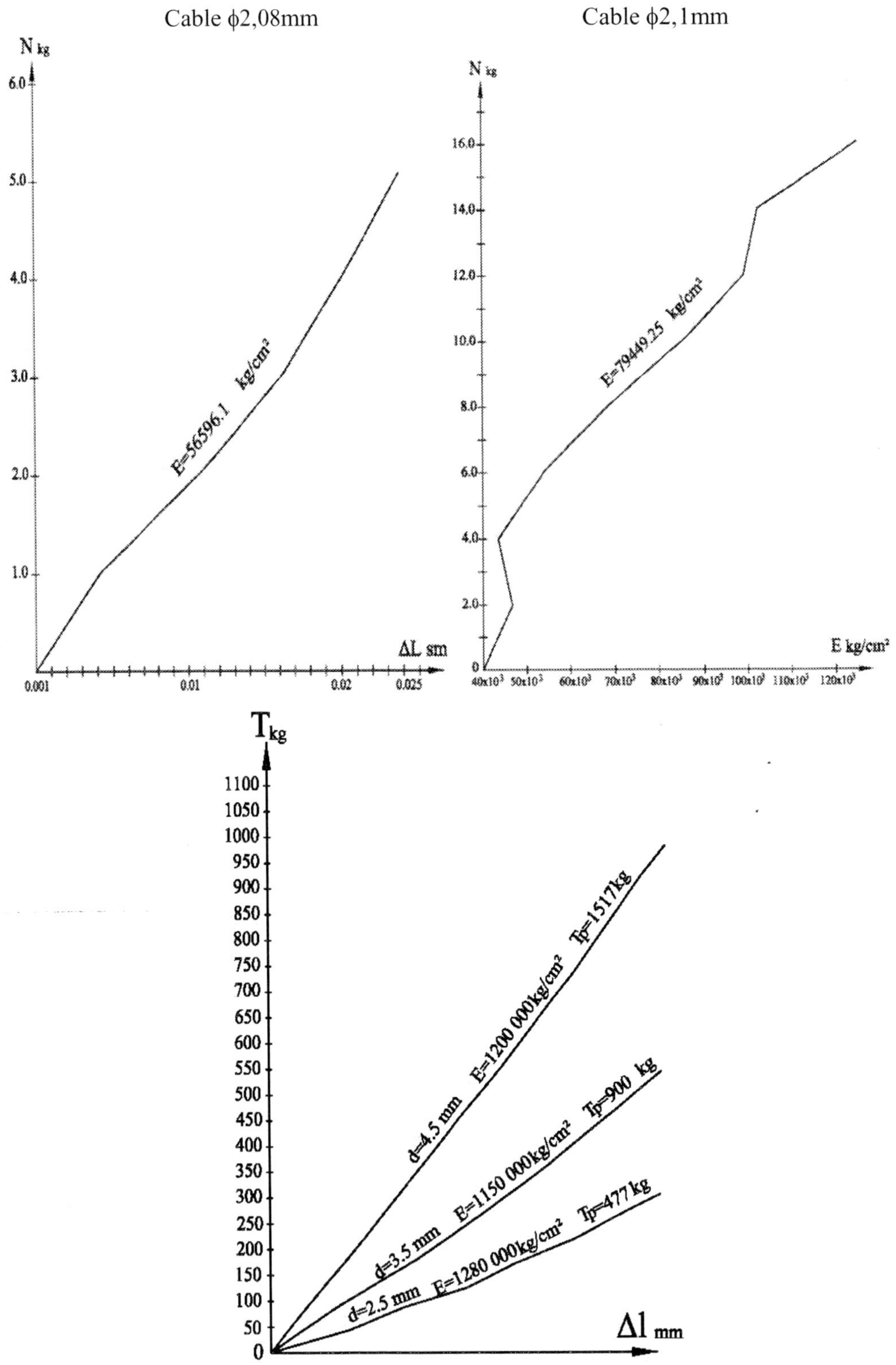

Figure 2.12. Determination of the modulus of elasticity of cables.

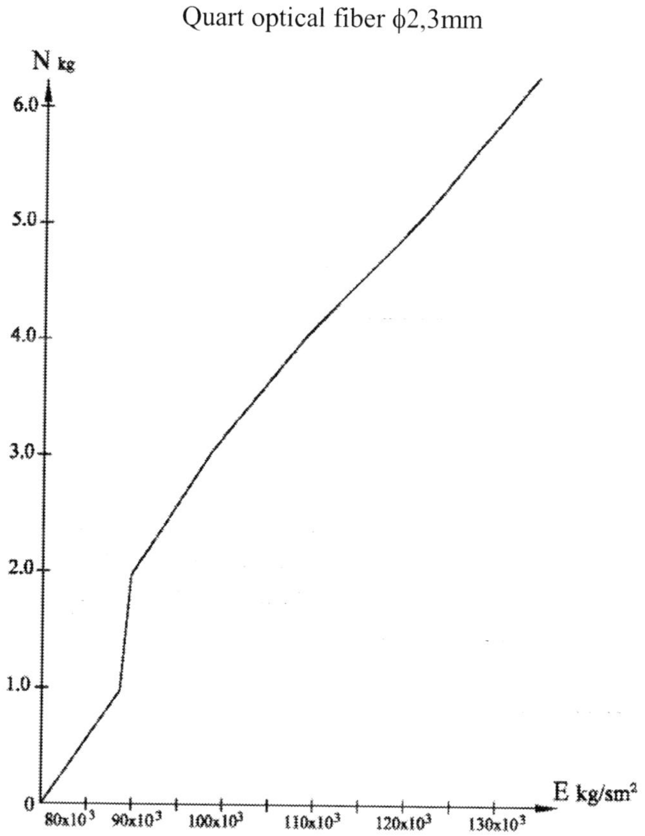

Figure 2.13. Determination of the modulus of elasticity of cables and fibers.

Figure 2.14. Experimental test stands.

When load achieved 5.0 kg optical fiber rod straightened and together with the cable was included in operation.

2.4. Estimation of the Sizes of Electromechanical and Fiber-Optical Sensors Clearings

The regulation of cable and guy constructions is connected with estimation of strain and respectively, elongation when electric or light signal from sensors is transmitted to drive switching system. These values denote: in the case of electromechanical sensor with $\Delta l_{\text{АЭСМ}}$ and in the case of fiber-optical sensor with $\Delta l_{\text{ВОСМ}}$. For each concrete composite construction $\Delta l_{\text{АЭСМ}}$ and $\Delta l_{\text{ВОСМ}}$ are to be defined which ensure swithing of drives, and stretching-compression of controlled cables. Below, some diagrams of the electromechanical and fiber-optical sensors are given.

As it is known curve length y=y(x) between two points $M_1(x_1,y_1)$ and $M_2(x_2,y_2)$ is calculated with formula:

$$S = \int_{x_1}^{x_2} \sqrt{1 + y^{12}(x)dx} \tag{2.45}$$

In the case of circle the curve length is expressed with formula:

$$S = \ell\sqrt{1 + \frac{16}{3}\left(\frac{f}{\ell}\right)^2} \tag{2.46}$$

In the case of parabola the length of curve is expressed with formula:

$$S = \ell\left[1 + \frac{8}{3}\left(\frac{f}{\ell}\right)^2\right] \tag{2.47}$$

In the case of sinusoid the length of curve is:

$$S = \int_0^{\ell}\left[1 + \frac{1}{2}\left(\frac{dy}{dx}\right)^2\right]dx \tag{2.4}$$

If

$$y(x) = f\sin\frac{n\pi x}{\ell};\; y' = f\frac{n\pi}{\ell}\cos\frac{n\pi x}{\ell}$$

$$(y')^2 = f^2\frac{n^2\pi^2}{\ell^2}\cos^2\frac{n\pi x}{\ell}$$

$$S = \int_0^\ell dx + \frac{1}{2}\int_0^\ell f^2 \frac{n^2\pi^2}{\ell^2}\cos^2\frac{n\pi x}{\ell}dx = \ell + \frac{1}{2}f^2\frac{n^2\pi^2}{\ell^2}\int_0^l \cos^2\frac{n\pi x}{\ell}dx =$$

$$= \ell + \frac{1}{2}f^2\frac{n^2\pi^2}{\ell^2}\left\{\left[\frac{1}{2}x\Big|_0^l + \frac{1}{4a}\sin 2\frac{n\pi x}{l}\Big|_0^l - \right]\right. =$$

$$= \ell + \frac{1}{2}f^2\frac{n^2\pi^2}{l^2}\left[\frac{1}{2}\ell + \frac{1}{4\frac{n\pi}{\ell}}\sin 2\frac{n\pi x}{\ell} - \frac{1}{4\frac{n\pi}{\ell}}\sin 0\right] = \tag{2.49}$$

$$= \ell + \frac{1}{4}f^2\frac{n^2\pi^2}{\ell^2}\ell;$$

$$S = \ell\left(1 + \frac{1}{4}f^2\frac{n^2\pi^2}{\ell^2}\right)$$

where n is quantity of waves.

Example

Span of optical fiber ℓ=10cm;

Optical fiber hoisting arm f=1,0cm;

Test length of the cable ℓ_{cab}=50cm;

Modulus of elasticity of the cable E_{cab}=56696kg/cm^2;

Cable cross-section area A_{cab}=0,03396cm^2 (∅ 2,08mm).

From Hook's law $\Delta\ell_{cab} = \dfrac{N\cdot\ell_{cab}}{A_{cab}\cdot E_{cab}}$.

Optical fiber gain after linearization:

$$\Delta\ell_{cab} = S - \ell$$

* For circular curve:

$$S_{cab} = 10\sqrt{1 + \frac{16}{3}\left(\frac{1}{10}\right)^2} = 10,263cm$$

$$\Delta\ell_{cab} = 10,263 - 10 = 0,263cm$$

Strain in cable $N = 0,263\cdot 0,03396\cdot 56696/50 = 10,13kg$

* For parabolic configuration:

$$S_{cab} = 10\left[1 + \frac{8}{3} \cdot 0,01\right] = 10,27cm$$

$$\Delta\ell_{cab} = 10,27 - 10 = 0,27cm$$

Strain in cable N=0,27·0,03396·56696/50=10,38kg

* For sinusoid configuration (one wave):

$$S_{cab} = 10\left[1 + \frac{3,14^2}{4}\left(\frac{1}{10}\right)^2\right] = 10,246cm$$

$$\Delta\ell_{cab} = 10,246 - 10 = 0,246cm$$

Strain in cable N=0,246·0,03396·56696/50=9,47kg which is practically close to test result.

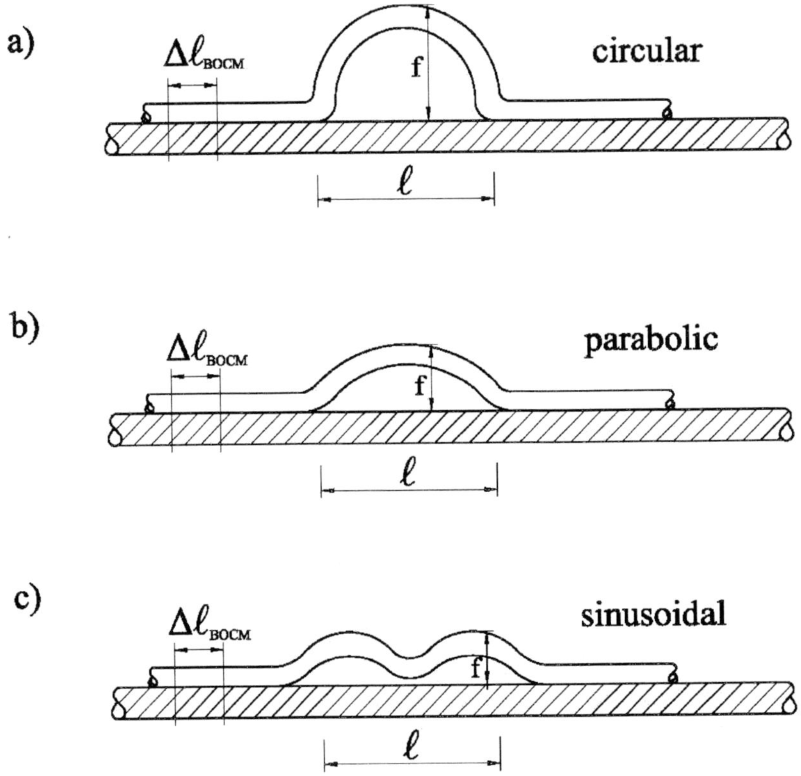

Figure 2.15. (Continued)

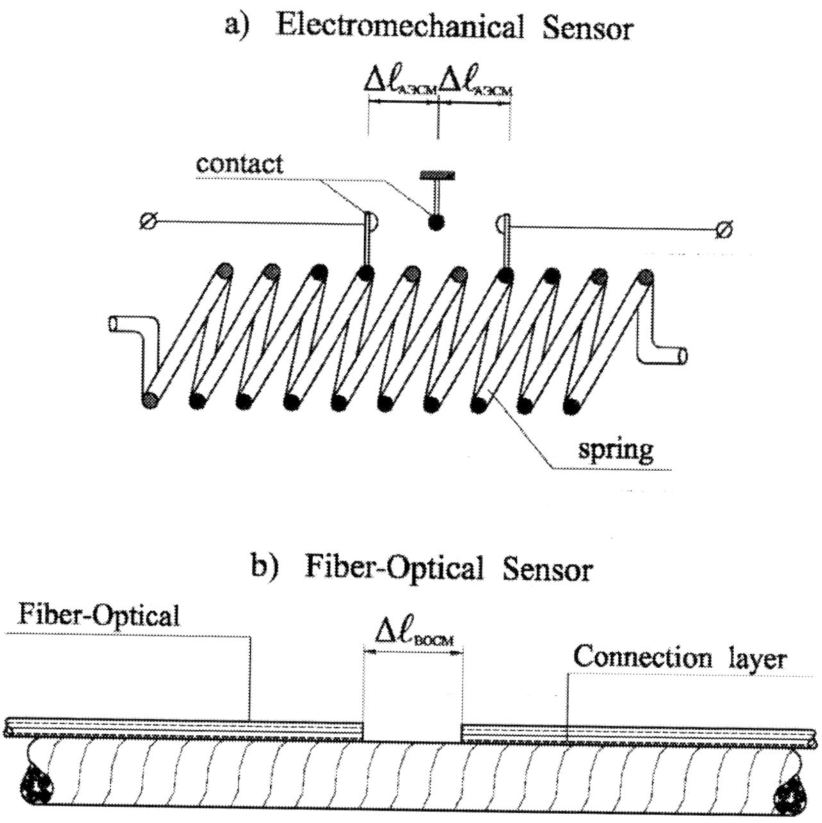

Figure 2.15. Optical fiber configurations: a) circular; b) parabolic: c) sinusoidal Diagrams for cleagap and gap. Definizions:
АЭСМ – automatic electromechanical sleeve with contactors' sensor;
БОСМ – automatic electric stretching sleeve with fiber-optical sensor.

2.5. Development of the Diagrams of Fiber-optical Sensors for Realization of the Corresponding Methods

Modification of schematic diagrams of fiber-optical systems for structures control-regulation and breaking prognosis is mainly based on operation principles of fiber-optical light guides, photo receivers, radiators and other components of optical diagram.

There are two types of light passing in fiber: meridional and indirect (Figure 2.16)

Consider the physical property of optical fiber which is more important when used as a sensitive element. Consider the character of light phase variation propagated along the fiber which changes under the action of mechanical pressure, temperature and compression forces. Phase variation (βl) presented with the measured object at using interferential fiber-optical sensor will become apparent as the variation of interference intensity. Thus, for determination of fiber-optical sensor sensitivity under increased pressure, temperature and so on, it is necessary to know phase variation (βl) relative to these values. Define light phase ψ of sensor fiber. At small comparative difference of refractory indices in the fiber:

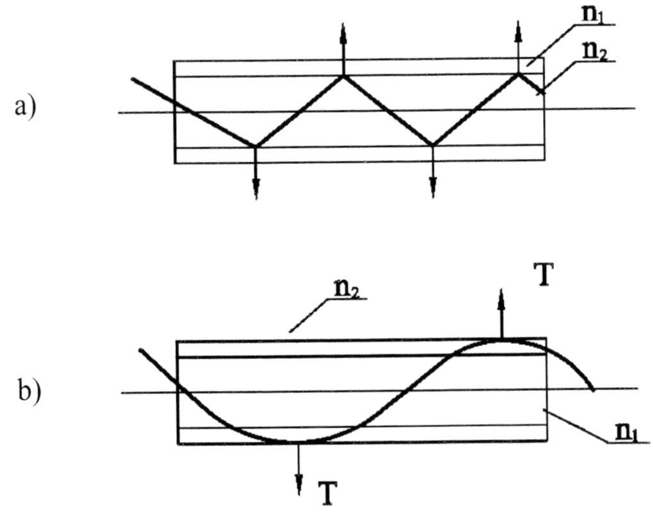

Figure 2.16. Types of light passing in fiber: a) meridional, b) indirect.

$$\psi = \beta l \approx Knl \tag{2.56}$$

where n is the refractory index of core.

Then comparative variation of phases under the action of measured object (sensitivity) is equal to:

$$\Delta\psi / \psi = \frac{\Delta l}{i} + \frac{\Delta n}{n} \tag{2.57}$$

According to the derived formula it is possible to make a comparatively sensitive sensor for measuring construction strength, humidity, pressure, temperature, etc. parameters.

Light radiation in a fiber is disperced on many uniform plane waves with space-time distribution of the determined field. Each such wave is characterised with self-constant $\beta = K_0 n_1 \cos\theta_n$, phase $V_\varphi = \omega / \beta n$ and group speed $V_{2n} = \delta\omega / \delta\beta_n$. The condition of phase matching among fallen and reflected waves is always fulfilled.

Consider the picture of beam passing in the fiber (Figure 2.16). Beams are divided into two types: meridional and indirect. Meridional beams cross the fiber axis, while the direction of indirect beams does not intersect with the axis.

On propagation of meridional beam the refraction angle θ is less than internal reflection critical angle θ<θ. This is expressed as:

$$\cos\theta_{cr} = \frac{n_2}{n_1} \tag{2.58}$$

All those beams the angle of incidence of which is less than angle θ will be arrested in the middle, while

$$\sin\theta = \sqrt{n_1^2 - n_2^2} = 4A \tag{2.59}$$

and value A is called numerical aperture of the fiber.

At propagation of indirect beams the falling of beam on the wall is determined with θ, φ and v angles in incidence point (Figure 2.16.a). Figure 2.16.b shows beam projection on the plane in the point of incidence on core border and on the plane of incidence. Plane of incidence makes angle φ with core axis, v fills up the angle of incidence:

$$\cos\theta = \cos v \cdot \cos\varphi \tag{2.60}$$

Line crossing the borders on the plane of incidence represents an ellipse. In the case of borders curving there happens full distortion of internal reflection and radiation leaking in the shell.

When θ<θ the meridional and indirect beams form the modes of directed core. When θ becomes more than θ, in the case of indirect beams there will happen reflection disturbance and this will continue until the angle v of propagation is less than angle θ. The corresponding critical value of angle θ is defined as:

$$\cos\theta_{\lim} = \cos\theta_{cr} \cdot \cos\varphi \tag{2.61}$$

Critical angle increases from value $\theta_{lim}=\theta_{cr}$ to value $\varphi = \theta_{\lim} = \frac{\pi}{2}$. The indirect beams when angles of incidence are $\theta_{lim}<\theta<\theta_{cr}$ and $\varphi = \frac{\pi}{2}$ form the waves which because of leakage on interface are characterized with considerable losses. The received modes are called outcoming or shell modes. When $\theta>\theta_{cr}$, the radiation passes in the space at determination of continuous spectrum of space modes.

According to magnetic field type the directed modes of optical fiber are divided into two groups: EH_{nm} and HE_{nm} modes. n is azumuth series of transverse field of mode, while m is radial series of transverse field of mode. For EH_{nm} mode longitudinal component magnetic field dominates over longitudinal electric field, while for HE_{nm} - vice versa. When $n=0$, axially symmetrical modes are formed.

The main parameter is constant propagation β. It is determined with phase and with group speed of modes. HE_{11} mode is called the main mode. At value $V<2,4$ in the fiber is realized a single mode regime – only main mode HE_{11} is propagated. With the increase of parameter V the possibility of propagation of higher order modes will arise. The borders of constant propagation are:

$$k_0 n_2 < \beta < k_0 n_1$$

Optical fibers with slightly differing refractive indices of core and jacket $n_1 \approx n_2$ represent weekly directing fibers. The advantage of such fibers is low distortion of signal and core dimensions more practical and useful increase. In week fibers every HE_{nm} mode is degenerated into mode EH_{n-2} of the same radial series. At the same time it is by two units lower to azimuth series of mode HE_{nm}. The field received with superposition of these modes is linearly polarized similar to initial field HE_{nm} and modes $E_{n-2,m}$. At the same time weekly directing fibers which are mainly used in optical contacts in interferometry and diagnostic systems are uniformly polarized in all directions of fiber cross-sections. For beam conducting in fibers lasers are generally used. The waves in output beam of laser have transverse elecromagnetic field with linear and uniform polarization.

Each value of parameter V of the fiber is corresponded with the respective modes m. At the same time, the values of constant propagation β correspond to the main mode. The value of β decreases with the increase of modes series to the value $\beta = k_0 n_2$.

Generally photoreceivers are of two kinds: single channel and multichannel. Multichannel photoreceivers, for their part are divided into two groups: linear and matrix ones. We have used multi-element matrix photoreceiver with charge connection which is distinguished for the simplicity of electronic mounting and control. In these matrices the elements can be summed up according the both coordinates, this enabling to create flexible hybrid metering schemes.

Matrix is a semiconductive device characterized with the complex of time parameters which fit well with time factors of dynamic processes and provide mechanical reaction conversion into signal.

Matrices allow the registration of nonstationary short time deformation processes. Particularly, enable to fix characteristic phenomena in 1 microsec.

Fixation of representation with single section matrix is realized in several microseconds. The formation of picture is done at signal pulse duration of 1-100 microseconds, its division into channels with fiber-optical light guides is done according to high intensive light source.

Consider fiber-optical sensors for construction control.

Figure 2.17 gives a schematic diagram of fiber-optical sensor. It consists of light source 1, fiber-optical light guides 2, light modulator 3 and photoreceivers 4. The operation principle is as follows: from light source 1 the beam via fiber-optical light guide 2 passes through light modulator 3 and via the second fiber-optical light guide 2 is transmitted to photoreceiver 4. The modulator is a thin elastic pipe, the compressve force acting on it causes the narrowing of its cross-section which, for its part, causes the change of light flux intencity. It represents a passing type fiber-optical sensor.

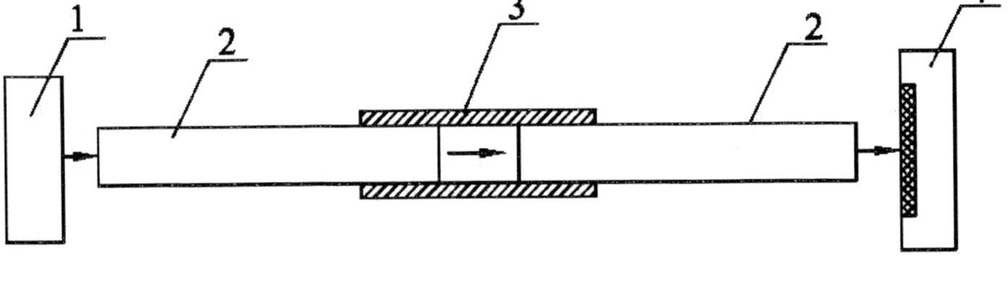

Figure 2.17. Passing type fiber-optical sensor.

From technical optics it is known that the flux passing in modulator or "light pipe" is equal to:

$$F = \frac{B\cos\varepsilon_1 \cos_2 S_1 S_2}{R^2} \tag{2.62}$$

Figure 2.18 gives the schematic diagram of fiber-optical sensor where monofiber-optical light guide sensitive on the whole length represents the modulator. It consists of light source 1, monofiber-optical light guide sensitive on the whole length 2 and photoreceiver 3.

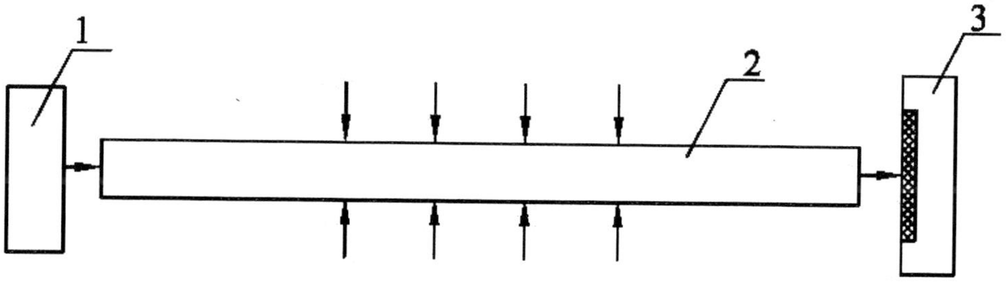

Figure 2.18. Fiber-optical transmitter with fiber sensitive on the whole length.

Consider in detail physical properties of a sensitive fiber.

The variation of glass refractive index is expressed with formula:

$$\frac{\Delta n}{n} = \frac{1}{n}\left(\frac{\delta n}{\delta T}\right)_\rho \Delta T + \frac{\delta n}{n} \tag{2.63}$$

where the first member takes into account the variation of glass density while the second member takes into account photo elasticity effect conditioned with fiber deformation, in particular, with lengthening or shortening because of pressure or temperature. For quartz glass:

$$\frac{1}{n}\left(\frac{\delta n}{\delta T}\right)_\rho = 0{,}68\cdot 10^{-5\circ} C \tag{2.64}$$

The variation of refractive index δn can be expressed with Pokels coefficient P_{ij} using the following formula:

$$\delta n = -\frac{n}{2}\left(P_{11}\varepsilon_1 + P_{12}\varepsilon_2 + P_{12}\varepsilon_z\right) \tag{2.65}$$

where ε_1 and ε_2 are relative deformations in cross-section and ε_z is deformation along the fiber axis, and:

$$\varepsilon_z = \frac{\Delta l}{l} \tag{2.66}$$

while

$$\varepsilon_1 = \varepsilon_2 = -\mu\varepsilon_z \tag{2.67}$$

When using these two formulas we can get:

$$\delta n = -\frac{n^3}{2}\left[\varepsilon_z\left(P_{12} - \mu\left(P_{11} + P_{12}\right)\right)\right] \tag{2.68}$$

If we substitute formula (2.63) into formula (2.) we get:

$$\frac{\Delta\psi}{\psi} = \frac{\Delta l}{l} + \frac{\Delta n}{n} = \varepsilon_z\left[1 - \frac{n^2}{2}\left(P_{12} - \mu\left(P_{11} + P_{12}\right)\right)\right] + \frac{1}{n}\left(\frac{\delta n}{\delta T}\right)_\rho \Delta T \tag{2.69}$$

Thus, in order to calculate the sensitivity of fiber-optic sensor it is necessary to determine deformation along the fiber caused by pressure, temperature and other effects. Let's turn to the dependence between variation of deformation, pressure and temperature. For simplicity assume that pressure *P* on the fiber acts uniformly and besides, acts on temperature ΔT variation. It is known that strain distribution inside coaxial cylinders is:

$$\sigma_v^{(i)} = A_i + \frac{B_i}{v^2},\quad \sigma_\theta^{(i)} = A_i - \frac{B_i}{v^2},\quad \sigma_z^{(i)} = C_i \tag{2.70}$$

where indices i=1, i=2 and i=3 denote data for (A) core, (B) jacket and (C) outer cover, respectively; while A_i, B_i and C_i are constants.

Strain inside core is final, therefore B_i=0, and besides, there are the following relations between strain (σ), deformation (ε) and displacement (u):

$$\left.\begin{aligned}
\varepsilon_v &= \frac{\partial U_v}{\partial v} = \frac{1}{E}\left[\sigma_v - v\left(\sigma_\theta + \sigma_z\right)\right] + \alpha\Delta T \\
\varepsilon_\theta &= \frac{U_v}{v} = \frac{1}{E}\left[\sigma_\theta - v\left(\sigma_z + \sigma_v\right)\right] + \alpha\Delta T \\
\varepsilon_z &= \frac{\partial U_z}{\partial z} = \frac{1}{E}\left[-\sigma_z - v\left(\sigma_v + \sigma_\theta\right)\right] + \alpha\Delta T
\end{aligned}\right\} \tag{2.71}$$

where *E* is Young module, α is thermal linear expansion coefficient; for quartz glass E=7750mgpa, $\alpha = 5{,}4\cdot10^{-7}K^{-1}$, *A, B* and C constants are defined from expressions:

$$\left.\begin{array}{ll}\sigma_z^{(1)}(a)=\sigma_z^{(2)}(a),\ \mathrm{U}_z^{(1)}(a)=U_z^{(2)}(a), & \mathrm{z=a}\\ \sigma_z^{(2)}(b)=\sigma_z^{(3)}(b),\ \mathrm{U}_z^{(2)}(b)=U_z^{(3)}(b), & \mathrm{z=b}\\ \sigma_z^{(3)}(x)=-p, & \mathrm{z=c}\end{array}\right\} \tag{2.72}$$

$$\left.\begin{array}{l}\int_a^{2\pi}\int_0^{a}\sigma_z^{(1)}vdvd\theta+\int_0^{2\pi}\int_{a0}^{b}\sigma_z^{(2)}vdvd\theta+\int_0^{2\pi}\int_b^{c}\sigma_z^{(3)}vdvd\theta=0\\ U_z^{(2)}=U_z^{(2)}=U_z^{(3)}\end{array}\right\} \tag{2.73}$$

Formulas (2.72) indicate that radial stress and displacement on boundary are continuous and also show the reletaion of outer forces. The first formula of (2.73) prove that both ends of the fiber are free, while the second formula shows that deformation on the ends are not considered as it is too small compared to surface deformation.

According to boundary conditions the constants A, B and C are defined with (2.71) and (2.73). If we substitute their values into formulas (2.70) and (2.71), we can find the dependence of deformation to pressure and temperature variation along fiber axis (Figure 2.19).

If we insert the values received by the above presented way in formula (2.3), then relative variation of phases $\Delta\psi/\psi$ can be determined.

Consider fiber-optical sensor where modulater is located at the end of fiber-optical light guide (Figure 2.20). The sensor consists of: light source 1, fiber-optical light guide 2, modulator 3, semitransparent mirror 4 and photoreceiver 5. This is a reflecting sensor. The principle of its operation is as follows: beam from light source passing through semitransparent mirror 4 and fiber-optical light guide 2 hits upon modulator 3. As a result of physical action on modulator light beam changes one parameter out of five, returns back through light guide 2 and semi- transparent mirror and finally falls on photoreceiver 5.

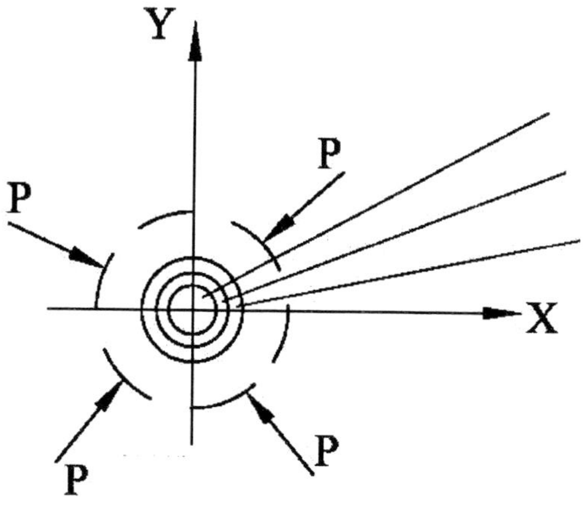

Figure 2.19. The diagram of fiber-optical light guide under the action of uniform pressure.

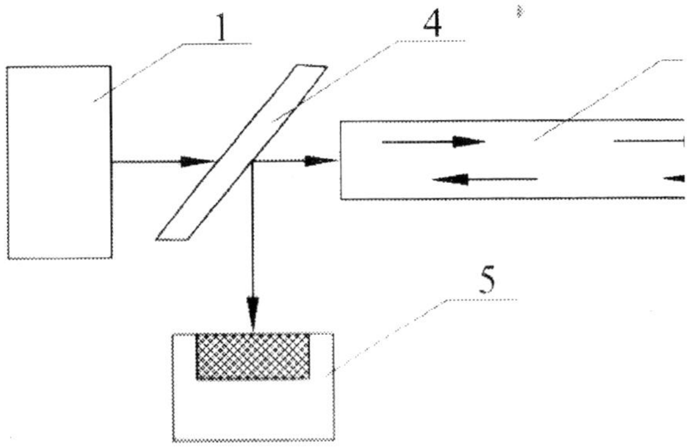

Figure 2.20. Reflecting type fiber-optical sensor.

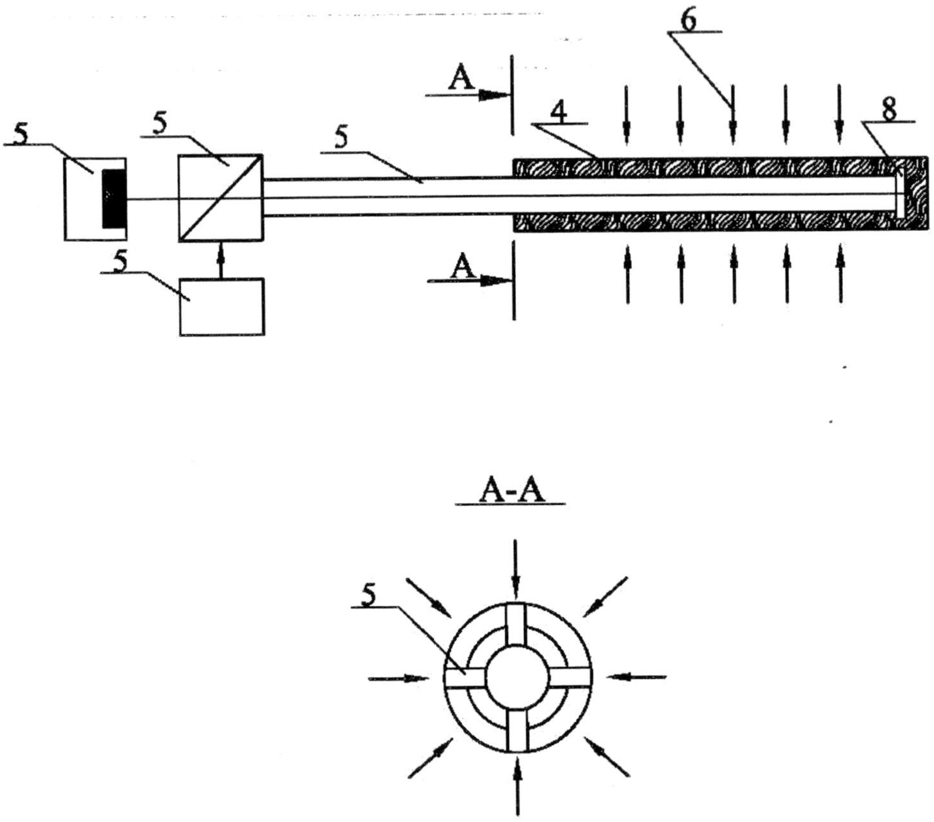

Figure 2.21. Fiber-optical sensor with profiled deformer.

Figure 2.21 presents a fiber-optical sensor where profiled cylinder performs the function of a modulator. It can be fixed in concrete construction. It consists of light source 1, autocolimation block 2, fiber-optical light guide 3, profiled cylinder 4 with flutes 5, photoreceiver 7 and reflecting mirror 8. External load 6 is acting on profiled cylinder 4. The sensor operates as follows: radiation from light source 1 through autocollimation system 2 is directed to light guide 3 where after multiple refracton, it falls on nontransparent mirror 6, the beam reflected from mirror returns back and by means of autocollimation system 2 falls on photodetector 7. When stress does not act on profiled deformer the signal value does not change on photodetector 7. When stress begins to act on cylindrical profiled deformer 4, the light guide 3 undergoes microbendings and signal value begins to change (decrease) on photodetector. If we connect this with light intensity force we shall receive force value acting on deformer.

Figure 2.21 shows functional diagram of the system of fiber-optical sensor diagnostics.

The proposed sensor enables to define continuously and precisely the forces in concrete construction and estimate its stress-strained state.

It consists of the following parts: test object, fiber-optical sensor, optical transmissions block, automatic control of functional systems, photoreceiver block, analog-to-digital transformer, filter, multiparametric analyzer, data base, neuron network, block and monitor generating alarm and emergency prognostication signals.

The proposed system operates as follows: light pulse from optical transmitter is fed to fiber-optical sensor which is located on the test object. Pulse passing the sensor gets to photo-receiver block. Synchronization of optical radiator and photoreceiver is regulated with automatic control functional systems. Signal from photoreceiver passes through filter, multiparameter analyzer and gets to data base where data are processed with mathematical models of Fourier spectral analysis. Then neuron network selects the data, those that are unimportant are discarded to waste-bin, and those giving prognosis are sent to block of formation of alarm and emergency prognostication signals. The results are desplayed on the monitor.

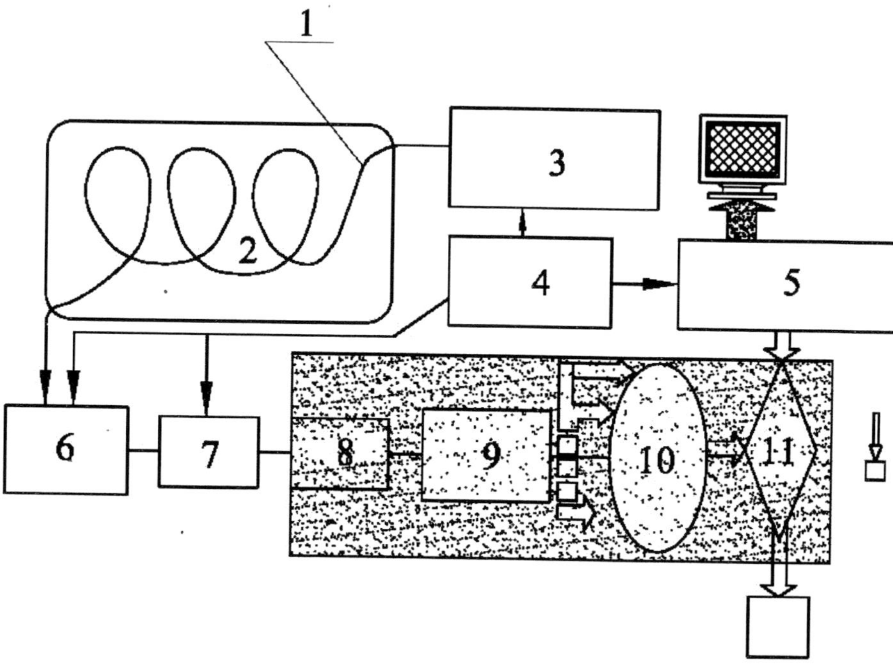

Figure 2.22. Functional diagram of the system of fiber-optical diagnostics.

2.6. Metrological Estimation of the Constructed Sensors

The quantitative measure of sensors measuring precision is their error. Aggregate error of a sensor is composed of many initial errors which depend on measuring method, diagram, quality of tool, character of proceeding of check process, environment conditions and other external or internal factors.

Sensor operation working on directional light guide in general is described with complex function of many parameters including equipment parameters, as well as, exploitation conditions characterizing parameters which change under the effect of external factors. With consideration of significant parameters acting on sensor operation, this function can be written as:

$$I = f\left(\alpha_0, n, \varphi, u, T\right) \tag{2.74}$$

where u is the stress of rediator power supply and T is temperature of environment.

The dependence between mean square and boundary Δ_{cr} errors (deviations) is defined with error distribution law Δ_{cr}=3 n.

Let's estimate the initial errors of the equipment. The radiation distribution angle is defined at equipment assembling. Assembling of sensor joints in a special fixture enables to ensure the angle α_0 with precision 0.05° which determines scattering of distribution angle θ_i.

The refraction angle of glass item may change. The degree of difference depends on glass type, on used spectrum, etc. Thus, for example, for glass TФ-10 at infra-red radiation the refractive index in one batch may differ as Δn=0,00097.

Refraction width of light diodes in one batch is varied, e.g. for light diodes АЛ170А dissipation makes about 10-15 %. Voltage is used as power sourse, the allowable deviation from nominal being not more than ±5 %.

The mean square error of a sensor is defined as:

$$\sigma_{I,} = \sqrt{\left(\partial I / \partial \theta_1\right)^2 \theta_1 = 45\sigma_2^2\theta_1 + \left(\partial I_n / \partial n_1\right) = 1{,}8\sigma_{n_1}^2 + \cdots + \left(\partial I \ / \partial T^2\right)} = 20$$

Let's determine partial derivatives in the points with normal conditions. Geometrically they represent slope tangents to output signals curve in corresponding points of sensors normal parameters.

The coefficient of radiation dissipation width effect is defined with formula $n_1 = K_n\left(\Phi\right) n_{2\max}$ for normal state. In order to determine the coefficient of the effect of precise setting of radiator on output signal we use the following expression:

$$\frac{\partial}{\partial \alpha_0} = \frac{\tau b_0 (I - I_0) S \pi n_1 \cos\left\{\pi / \arcsin\left[(n_1/n_0) \sin \arcsin n_2/n_1 - \alpha_0\right] \Phi\right\}}{4\Phi n_0 / 1 - \left[(n_1/n_0)\sin\arcsin n_2/n_1 - \alpha_0\right]^2} - \cos\left(\arcsin n_2/n_1 - \alpha_0\right) \tag{2.75}$$

The index of the effect of light guide material refraction dissipation on output signal of the sensor is:

$$\frac{\partial}{\partial \alpha_0} = \frac{\cos\left(\arcsin n_2 / n_1 - \alpha_0\right)}{\sqrt{n_1 / n_2 - 2}} - \sin\left(\arcsin n_2 / n_1 - \alpha_0\right) \tag{2.76}$$

The coefficient of supply source instability effect is determined as:

$$\partial I \;/\; \partial U \; = \tau \; b_0 s \;\; \cdot \tau\left(n_2, \Phi \;\right) I \;\; / \Phi \tag{2.77}$$

where Φ is thermal potential.

The correction of output signal of the sensor with experimentally received temperature coefficient considers temperature instability of light source, photoreceiver, light guide structure and controlled coating.

For normal environment the value of experimentally obtained coefficient at temperature instability is 0.08 microP/K, which corresponds to additional temperature error of about 1%K.

For radiation power stabilization optical feedback is used which enables to decrease instability to 0.15%K i.e.almost twice.

In order to increase the equipment precision, maximum frequent compensation or irradication of errors is necessary. The mentioned errors for sensors are the errors of dissipation width of radiator guiding diagrams, temperature errors and errors caused by instability of supply source.

2.7. Tests for Investigation of Stress-Strained State of Simple Elements with Fiber-optical Sensors

We designed and manufactured a test bench given in Figure 2.23. It consists of test object 1, fiber-optical sensor 2, fiber-optical accelerometer 3 and information processing block 4.

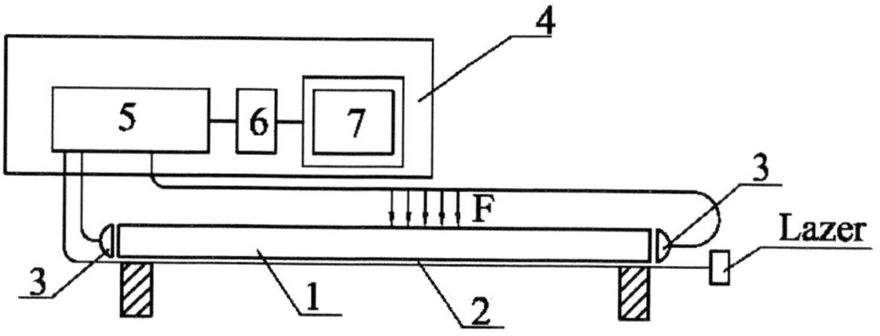

Figure 2.23. Test bench.
5-receiver-detector, 6-analog-to-digital transformer, 7-monitor.

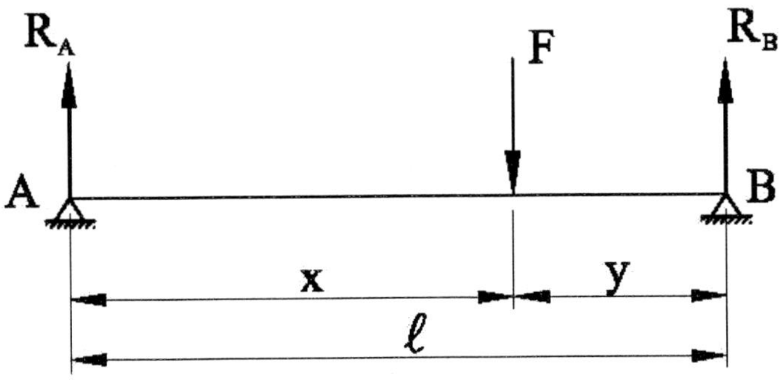

Figure 2.24. Calculation diagram.

By means of fiber-optical sensor which is attached on the whole length of element, we can determined its sag (deformation) and the value of acting strength. Reaction forces in supports are determined with accelerometer. If we determine reaction forces it is possible to determine the value of resultant force and point of engagement, i.e. the expected dangerous section. By this diagram for stating this section no reflectometer not needed. By mathematical model developed by us and contained in computer we can determine the precondition of element failure.

Here is given one example of the performed tests.

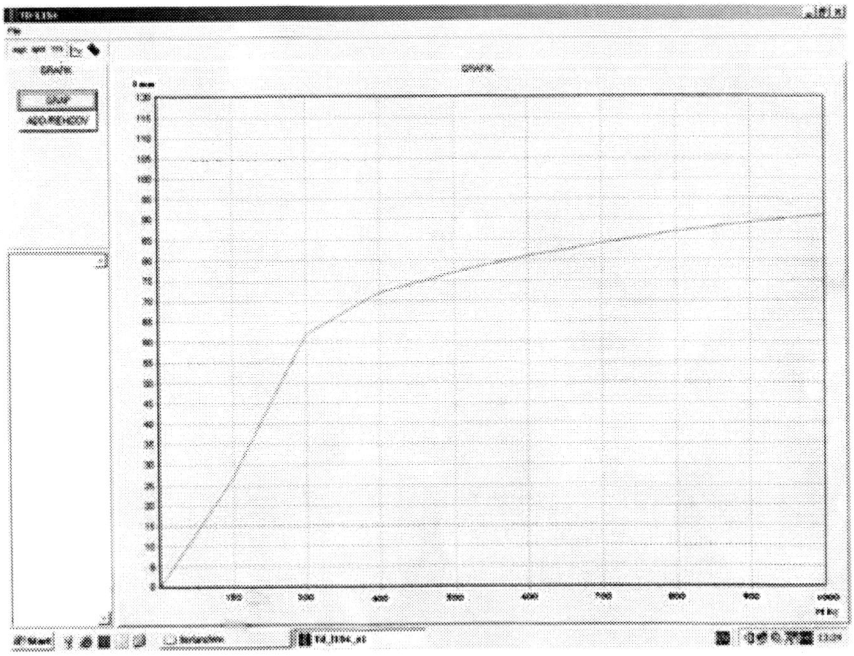

Figure 2.25. Results of measurements.

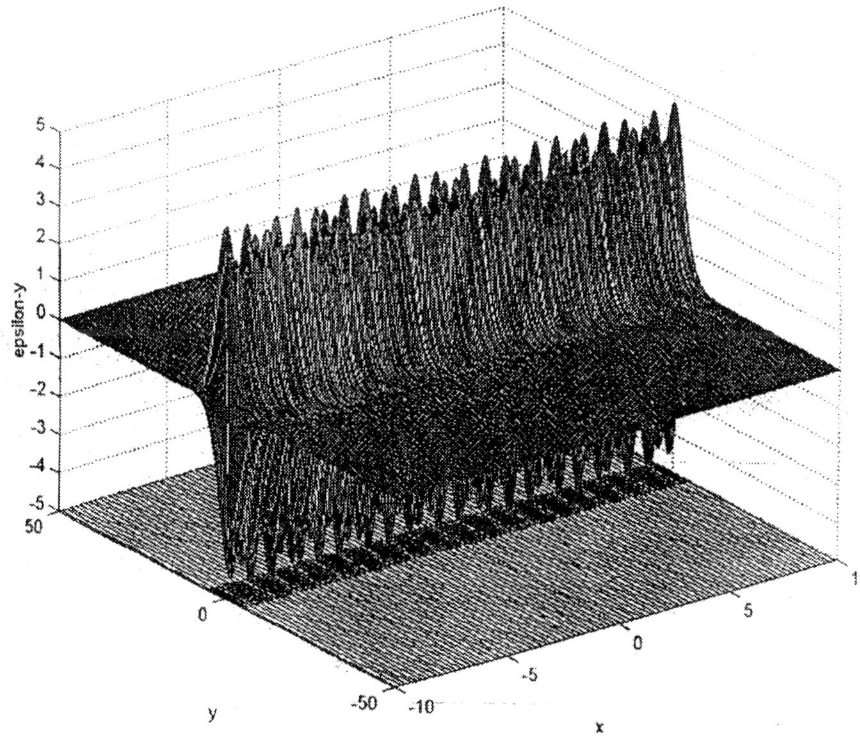

Figure 2.26. Fourier spectrum.

Example. On steel girder lying on two supports 3 a=10mm, b=20mm, l=3000mm, E=210mgpa, on which force F is acting (Figure 2.24).The point of force engagement, sag and force value, failure precondition and critical force are to be determined.

For the solution of the problem fiber-optical sensor is attached from below on the whole surface of the girder and accelerometer is mounted near the supports for determination of reaction forces.

The results of measurements are given in Figure 2.24 which shows that R_A and R_B are determined with accelerometers, after that according to the composed program X and Y distances or force F application points are determined, while acting force F is defined with fiber-optical sensors. Then force F is increased until girder failure which is registered with fiber-optical sensor.

The obtained signals were analyzed by Fourier spectral analysis given in Figure 2.26.

Fourier spectral analysis shows that peaks are consentrated in the section of maximum forces application and their amplitude reaches maximum that indicates the precondition of girder falure.

Here the deformations may also be determined. If before loading of construction the device reading is C_1, at the end of loading - C_2 and after unloading – C_3 then:

Absolute deformation $f=C_2-C_1$;
Residual deformation $f=C_3-C_1$;
Bending deformation $f=C_2-C_3$.

Chapter 3

Regulation of Operation of Combined Framed Structures Using Electromechanical and Fiber-optic Sensors

3.1. Introduction

There are several types of frame straining devices including those using hoist and planetary reduction gears. The diagrams of their application are given in Figure 3.1 and 3.2. In the carried out experiments the planetary reducers were used.

3.2. Automatic Regulation of Seismic Load for Metal Frames

Theoretical and experimental researches were executed on a frame model (see Figure 3.3).

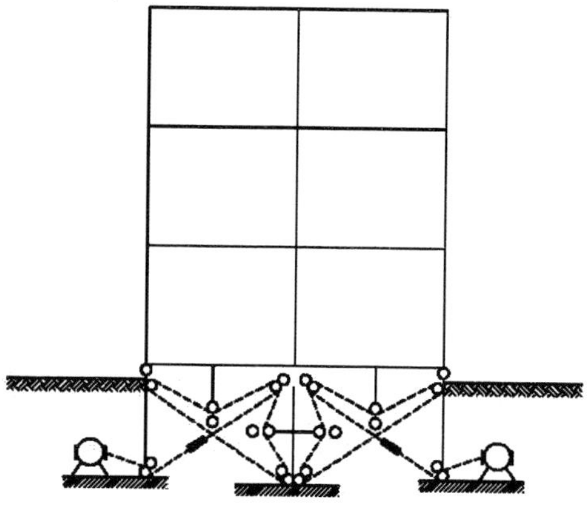

Figure 3.1. Diagram of hoist usage.

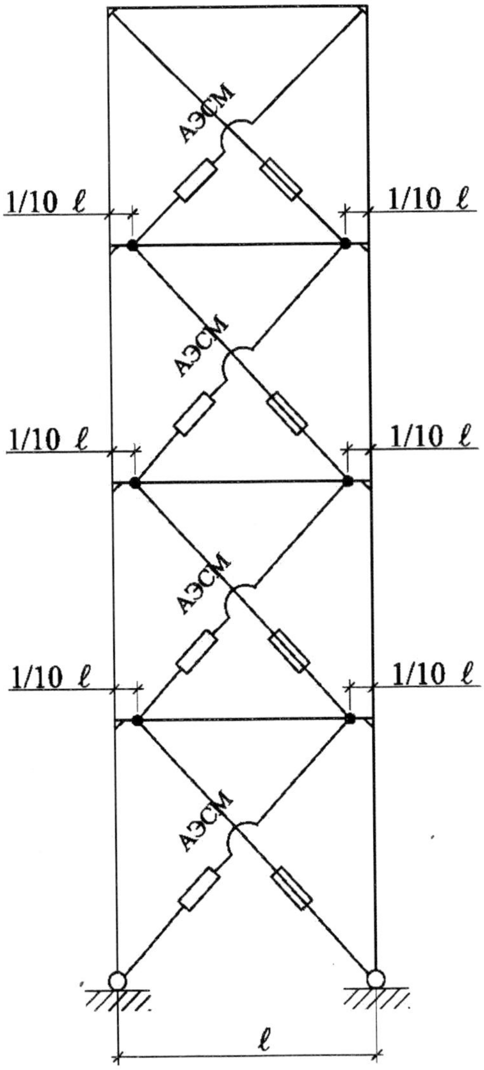

Figure 3 2. Diagram of using of planetary reducer.

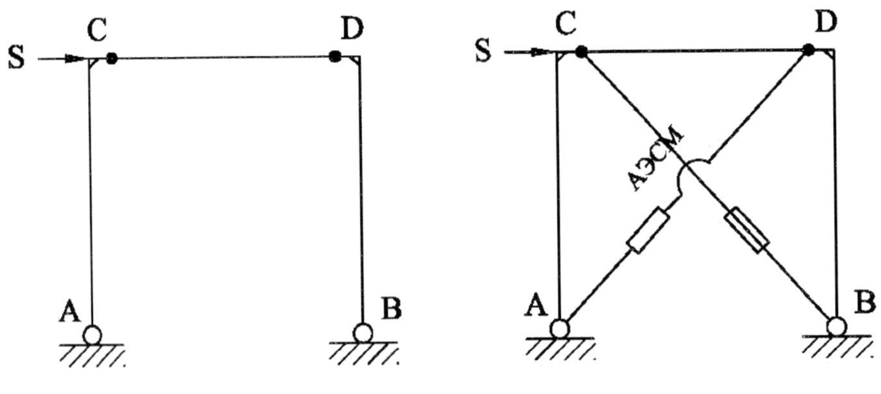

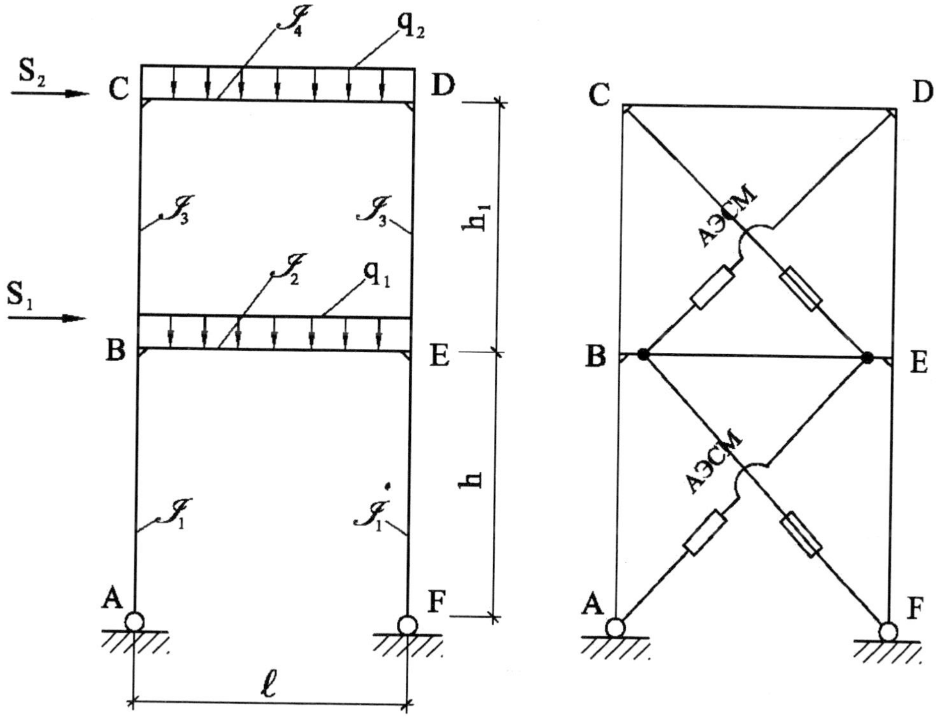

Figure 3.3. Frame models used in researches.

Model one. Single span one storey open contour frame was made of pipes. Pipe diameter was 32.5 mm, frame span ℓ= 680 mm, height h = 1240 mm. On one diagonal tie-bar with diameter 2.2 mm a planetary reducer with electric drive and a relay with contactors were mounted.

The second model had the same geometrical dimensions and cross-sections, except the lower collar beam support added in the distance of 150 mm which made closed contour of the frame (Figure 3.4).

As it is known horizontal seismic force S_{ik} is determined with the product of many coefficients including dynamic coefficient β_i which depends on the period of frame natural vibration and on its variation. Seismic loads may be regulated 2 ÷ 2.5 times.

In order to include АЭСМ into frame operation it is necessary to preliminary define gap between AECM relay contactors which is determined with formula:

$$\Delta\ell_{АЭСМ} = (T + \Delta T)\left(\frac{\ell_{\text{cont}}}{E_{\text{cont}} F_{\text{cont}}} - \frac{\ell_{\text{cobl}}}{E_{\text{cobl}} F_{\text{cobl}}}\right) \tag{3.1}$$

or

$$\Delta\ell_{АЭСМ} = \frac{S_{ik}}{K_i}\left(i_{\text{cont}} - i_{\text{cobl}}\right);\ \text{k}_i = \frac{S_{ik}}{(T + \Delta T)} \tag{3.2}$$

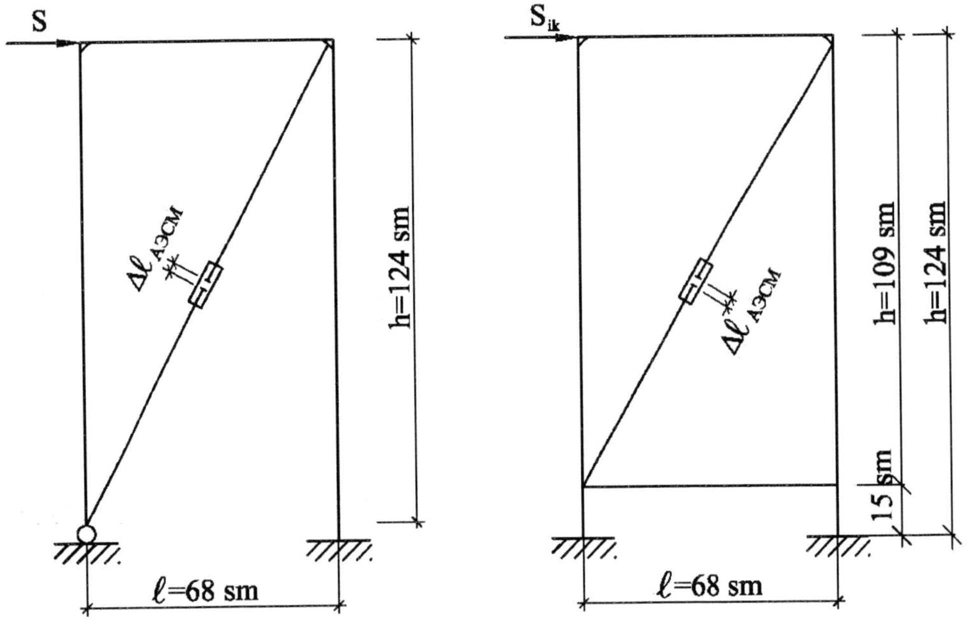

Figure 3.4. Closed contour frame.

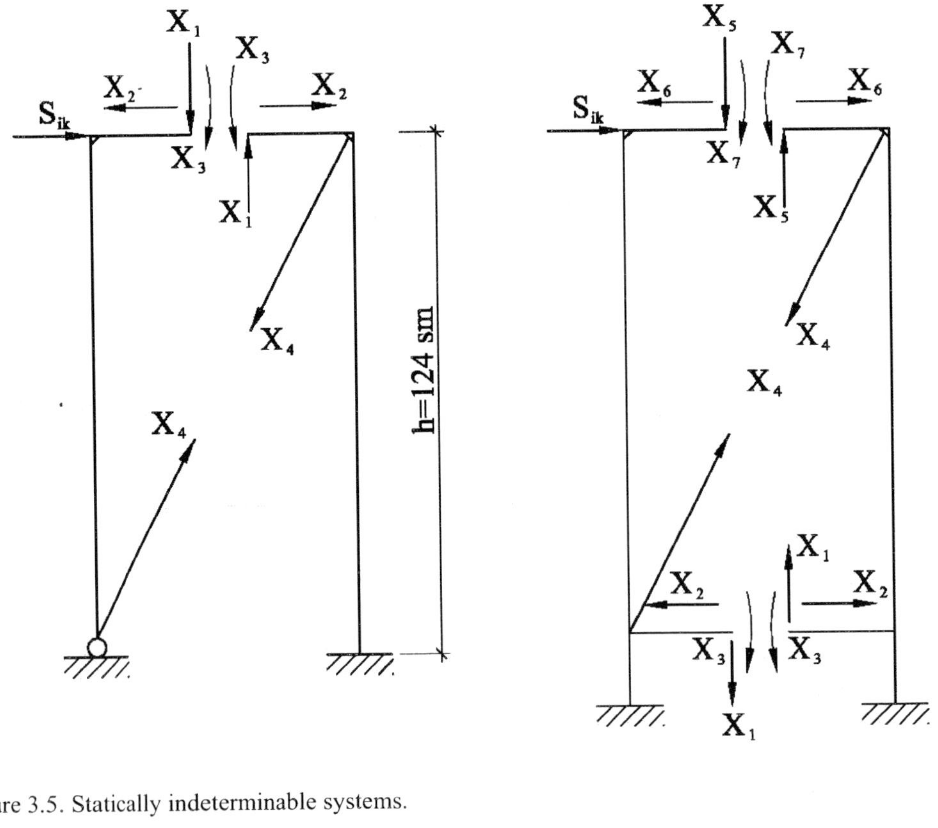

Figure 3.5. Statically indeterminable systems.

where T is prestrain in cable (in diagonal bracing);

S_{ik} is seismic force;

ΔT is cable (diagonal bracing) selfstressing force;

l_{cont} is the length of contactor spring and cable including AECM;

$E_{cont} F_{cont}$ is the stiffness of contactor spring and cable including AECM;

ℓ_{cabl} is cable (diagonal bracing) length without AECM;

$E_{cabl} F_{cabl}$ is cable (diagonal bracing) striffness without AЭCM.

Here E_{cont}, E_{cabl}, F_{cont}, and F_{cabl} are respective moduli of elasticity and cross-section areas of contactor spring and cable.

For determination of coefficient K_i three calculations and tests have been carried out.

The first calculation is based on classical method of forces according to which the mentioned frame models represent, in one case, 4 times statically indeterminable and, in the second case, 7 times staticaly indeterminable systems (Figure 3.5).

As a result of testing an open contour frame we get that when strain in diagonal bracing cable was T+ ΔT = 29.3 kg (0.293 kN), in case of horizontal static load S_{ik} = 175 kg (1.75 kN), the upper joint of frame collar-beam was displaced for f = 23.6 mm. After switching in of AECM displacement made f = 24.6 mm, strain in cable decreased and became T+ΔT=16 kg (0.16 kN). In this case coefficient K_i = 5.97.

Three calculations were done for the second, closed contour frame by method, as was mentioned above, of forces using program "SAP-2000 Student" and program "LIRA".

The canonical equation in matrix expression will be:

$$X_i \cdot \Delta = \vec{\Delta}_p \tag{3.3}$$

The equation can be written in the following form when S_{ik} =1:

$$i = \begin{Vmatrix} x_1 \\ x_2 \\ x_3 \\ x_4 \\ x_5 \\ x_6 \\ x_7 \end{Vmatrix}; \ \Delta = \begin{Vmatrix} 0,061 & 0 & 0 & -0,027 & 0,035 & 0 & 0 \\ 0 & 0,002 & -0,023 & -0,007 & 0 & 0,027 & -0,023 \\ 0 & -0,023 & 0,980 & 0,089 & 0 & -0,349 & 0,300 \\ -0,027 & -0,007 & 0,089 & 0,147 & -0,124 & -0,312 & 0,374 \\ 0,035 & 0 & 0 & -0,124 & 0,313 & 0 & 0 \\ 0 & 0,027 & -0,349 & -0,312 & 0 & 1,271 & -1,538 \\ 0 & -0,023 & 0,300 & 0,374 & 0 & -1,538 & 3,160 \end{Vmatrix}; \vec{\Delta}_p = \begin{Vmatrix} -0,059 \\ 0,013 \\ -0,175 \\ -0,006 \\ -0,261 \\ 0,636 \\ -0,769 \end{Vmatrix} \tag{3.4}$$

Square matrix of order 7 is reduced to that of order 5 and is solved using "MATCAD".

$$A := \begin{pmatrix} 0,002 & -0,023 & -0,007 & 0,027 & -0,023 \\ -0,023 & -0,98 & 0,089 & -0,349 & 0,300 \\ -0,022 & 0,274 & 0,308 & -0,974 & 1,153 \\ 0,027 & -0,349 & -0,312 & 1,271 & -1,538 \\ -0,023 & 0,300 & 0,374 & -1,538 & 3,160 \end{pmatrix};$$

$$b := \begin{pmatrix} 0,013 \\ -0,175 \\ -0,326 \\ 0,636 \\ -0,769 \end{pmatrix}; \; \text{lsolve}(A,b) = \begin{pmatrix} 255,08X10^{-3} \\ 3,31X10^{-3} \\ 2,35X10 \\ 1,08X10 \\ 5,28X10^{-3} \end{pmatrix} \qquad (3.5)$$

In experiments carried out for stating the relation of horizontal force S_{ik} and of tension $T+\Delta T$ in diagonal cable, S_{ik} = 50, 100 and 150 kg (0.5, 1.0 and 1.5 kN), as well as, T = 0, 20, 30, 40 and 50 kg (0, 0.2, 0.3, 0.4 and 0.5 kN) were varied. Their relation is expressed by coefficient K_i given in Table 3.1.

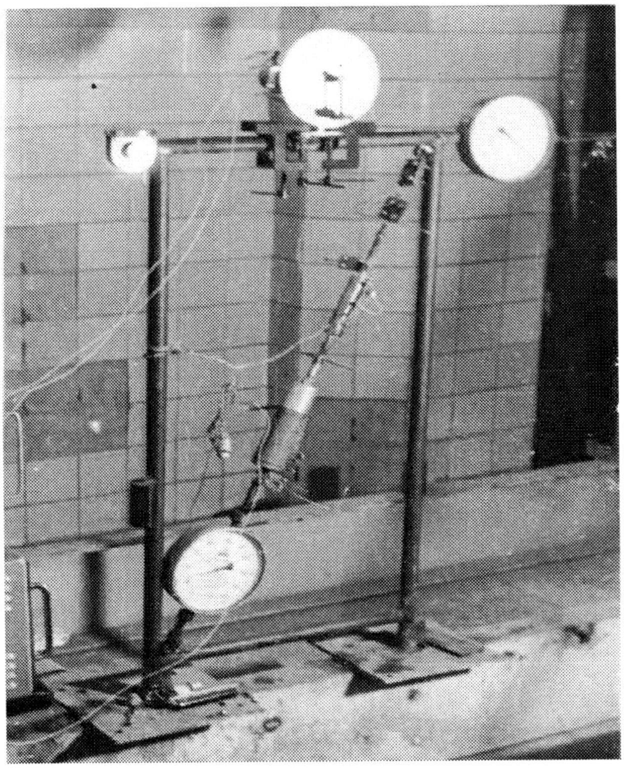

Figure 3.6. View of experimental steel frame with diagonal bracing.

Table 3.1. Coefficient $K_i = S_{ik}/(T+\Delta T)$

Research method	S_{ik} kg $T+\Delta T$	1 kg	50 kg	100 kg	150 kg	K_1	K_2	K_3	K_4	K average	Note
Method of forces	$T+\Delta T$	2,33kg	117,4kg	234,9kg	352,3kg	0,43	0,43	0,43	0,43	0.735	K_1 K_2 K_3 K_4 are numeration acoording S_{ik} loads In determination of K average value the result of test in case of preunstrained cable is not considered, T = 0
SAP-2000 Student	$T+\Delta T$	1,24kg	60,0kg	130,0kg	190,0kg	0,80	0,83	0,77	0,53		
"LIRA"	$T+\Delta T$	2,0kg	90kg	180,0kg	270,0kg	0,50	0,55	0,55	0,55		
Test ($T+\Delta T$) kg	T=0	-	10,0kg	12,0kg	18,0kg	-	5,0	8,33	8,33		
	T=20kg	-	44,0kg	82,0kg	127,0kg	-	1,13	1,22	1,18		
	T=30kg	-	62,0kg	104,0kg	146,0kg	-	0,80	0,96	1,03		
	T=40kg	-	78,0kg	122,0kg	163,0kg	-	0,64	0,82	0,92		
	T=50kg	-	88,0kg	136,0kg	180,0kg	-	0,57	0,74	0,83		

As is seen from the Table the relation between external horizontal force S_{ik} and tension in diagonal cable is ambiguous and is closer to the results of calculations when pretension in the cable is higher. $(T + \Delta T) = S_{ik}/0.735 = 1.36\ S_{ik}$ can be taken as an average value necessary for stating the initial magnitude ($\Delta\ell_{АЭСМ}$) of the gap between contactors.

Dynemic load in frame is induced with electric engine. Before switching in of automatic tie-coupling (AECM) the vibration amplitude was 1.5 mm, after switching in of AECM it was 0.5 mm.

In this case the efficiency of automatic straining tie-coupling was $n_{eff} = 3.0$; gap between contactors was $\Delta\ell_{АЭСМ} = 10$ mm.

3.3. Vibration Damping in Frames Using Moon Beam

3.3.1. Equation of Moon Beam Movement

Cantilever rod suspended on frame collar-beam, as a result of magnetic pull in the middle of span is stretched, bended and deviated from vertical.

As a result of magnets attraction the stretching force in the bar according to Hook's law is:

$$N = \frac{\Delta l}{l} EF \tag{3.6}$$

The length of the bar arc deviated from vertical is defined with expression:

$$L = \int_0^1 \sqrt{1 + \left(\frac{dy}{dx}\right)^2}\, dx \tag{3.7}$$

Hence bar elongation equals:

$$\Delta l = L - l = \frac{1}{2} \int_0^\ell \left(\frac{dy}{dx}\right)^2 dx \tag{3.8}$$

Then stretching force in the bar will get the following expression:

$$N = \frac{EF}{l} \cdot \frac{1}{2} \int_0^\ell \left(\frac{dy}{dx}\right)^2 dx \tag{3.9}$$

The static equation of stretched-bended bar will be:

$$EI \cdot \frac{d^4 y}{dx^4} - N\frac{d^2 y}{dx^2} = q \tag{3.10}$$

According to d'Alambert principle the bar movement equation may be written as:

$$m\frac{\partial^2 y}{dt^2} + EI \cdot \frac{d^4 y}{dx^4} - N\frac{d^2 y}{dx^2} = B(x,t) \tag{3.11}$$

Here the notations are:
m is linear mass of the bar;
EI is bar stiffness;
B(x,t) is magnets effect on the bar;
EF is bar tensional stiffness;
l - is bar length.
The solution of the equation can be presented as:

$$y(x,t) = W(x)T(t) \tag{3.12}$$

Substituting into the equation we get:

$$EI\frac{d^4W}{dx^4}Wdx \cdot T(t) + m\int_0^\ell Wdx\frac{d^2T(t)}{dt^2} - \frac{1}{2}\frac{EF}{\ell}\int_0^\ell\left[\frac{dW}{dx}T(t)^2 dx\right] \times$$
$$\times\frac{d^2W}{dx^2}Wdx = B(x,t)\int_0^\ell Wdx \tag{3.13}$$

Introduce notations:

$$\tilde{R}_{1x} = EI\int_0^\ell \frac{d^4W}{dx^4}Wdx; \quad R_{2x} = EI\int_0^\ell \frac{d^2W}{dx^2}Wdx; \quad R_{7x} = \int_0^\ell Wdx \tag{3.14}$$

Then motion equation for arbitrary exteral load will have the form:

$$T(t)R_{1x} + \frac{d^2T(t)}{dt^2}R_{5x} - \frac{EF}{2\ell}\tilde{R}_{1x}R_{2x}T^3(t) = B(x,t)R_{7x} \tag{3.15}$$

Consider magnets effect on the bar.
Unit length of the beam is acted upon with:

$$q = F_1 - F_2 = k_1\Phi_0^2/(a-y)^2 - k_1\Phi_0^2/(a+y)^2 \tag{3.16}$$

or

$$q = k_1\Phi_0^2/\left(a^2-2ay+y^2\right) - k_1\Phi_0^2/\left(a^2+2ay+y^2\right) = k_1\Phi_0^2\frac{\left(a^2-2ay+y^2-a^2-2ay-y^2\right)}{\left(a^2-2ay+y^2\right)\left(a^2+2ay+y^2\right)} =$$

$$= \frac{4a^2yk_1\Phi_0^2}{\left(a^2-2ay+y^2\right)\left(a^2+2ay+y^2\right)} \tag{3.17}$$

At small vibrations we have:

$$q = \frac{4k_1\Phi_0^2}{a^3}y \tag{3.18}$$

Static uniform magnetic field with constant stress of magnetic flow ϕ_0 is acting on the bar.

At bar deviation from vertical line the forces that act on it are caused by different magnetic fields and their difference does not equal zero.

Magnet attraction force between N and S poles is directly proportional to the square of magnetic flux ϕ_0 and inversely proportional to the square of distance between magnets.

$$\Delta F = F_2 - F_1 \quad F_1 = k_1\Phi_0^2/\left(a-y\right)^2; \quad F_2 = k_1\Phi_0^2/\left(a-y\right)^2 \tag{3.19}$$

or

$$F_{1,2} = k_1\Phi_0^2/\left(a \mp y\right)^2 \tag{3.20}$$

Attrective forces acting on the bar placed between magnets are intercompensated.

In the right part of vibration equation substitute $B(x,t)R_7$ with

$$q = \frac{4k_1\Phi_0^2}{a^3}y.$$

We get:

$$m\frac{\partial^2 y}{dt^2} + EI\cdot\frac{d^4y}{dx^4} - N\frac{d^2y}{dx^2} - \frac{4k_1\Phi_0^2}{a^3}y = 0 \tag{3.21}$$

Find the solution in the following form:

$$y(x,y) = W(x)T(t) \tag{3.22}$$

Inserting in equation and using Bubnov-Galerkin method we get:

$$\frac{d^2T(t)}{dt^2}\int_0^\ell mW^2dx + T(t)\int_0^\ell E\Im\frac{d^4W}{dx^4} - T(t)\int_0^\ell N\frac{d^2W}{dx^2} - T(t)\int_0^\ell N\frac{4k_1\Phi_0^2}{a^3}W^2dx = 0 \tag{3.23}$$

Inserting y(x,t) into $N = \frac{EF}{l}\cdot\frac{1}{2}\int_0^\ell\left(\frac{dy}{dx}\right)^2 dx$ we get:

$$N = \frac{EF}{l}\cdot\frac{1}{2}\int_0^\ell\left(\frac{dw}{dx}\right)^2 T^2(t)dx \tag{3.24}$$

Then

$$\frac{d^2T(t)}{dt^2}R_5 - T(t)\left[\int_0^\ell\frac{4k\Phi_0^2}{a_1^3}W^2dx - \int_0^\ell EI\frac{d^4W}{dx^4}W\,dx\right] - T(t)T^2(t)\times$$
$$\times\int_0^\ell\left[\frac{1}{2}\frac{EF}{l}\int_0^\ell\left(\frac{dW}{dx}\right)^2 dx\right]\frac{d^2W}{dx^2}Wdx = 0 \tag{3.25}$$

or

$$\frac{d^2T(t)}{dt^2} - \omega^2T(t) + \gamma T^3(t) = 0$$

We get Duffing equation where:

$$\omega^2 = \frac{\left[\int_0^\ell\frac{4k\Phi_0^2}{a_1^3}W^2dx - \int_0^\ell EI\frac{d^4W}{dx^4}W\,dx\right]}{\int_0^\ell mw^2(x)dx}$$

$$\gamma = \frac{\int_0^\ell\left[\frac{1}{2}\frac{EF}{l}\int_0^\ell\left(\frac{dW}{dx}\right)^2 dx\right]\frac{d^2W}{dx^2}Wdx}{\int_0^\ell mW^2(x)dx}$$

3.3.2. Magnetic Device of Moon Beam

Forced vibrations of bended cantilevel beam located in the strong field of two magnets can be adequately described with Duffing nonlinear differential equation.

For Moon beam operation a magnetic device is necessary. The suspended Moon rod is placed between two magnetic beams.

Tractive force of constant electromagnet is:

$$P_3 = \frac{(\bar{I}N)^2 4\pi 10^{-7}}{2} \frac{S_\delta}{\delta^2}(N) \qquad (3.26)$$

Gap between two electromagnetic beams is: δ=0,5m=50sm;

Design tractive force is: P=1t=1000kg=10000N;

Coil number on beam winding is: N = 2353 coils;

$\bar{I}$ is total current in winding.

Hence:

$$\bar{I} = \frac{2P_3\delta^2}{4\pi 10^{-7}\omega^2 S_\delta} \text{A} \qquad (3.27)$$

S_δ is gap area – electromagnet cross-section;

Wires of 20.8 layers on the beam with 112,6 coils in each layer make approximately 4700 m.

Beam height is 112.6 x 4.44 = 499.9 = 500 mm.

A projection of 2 x 250 = 500 mm is added to beam height.

The total height of the beam is 1000 mm.

We received 1500 mm. Minu suspended magnetic bar is placed in partition of double layer lanels, electromagnets are arranged in basement (Figure 3.7).

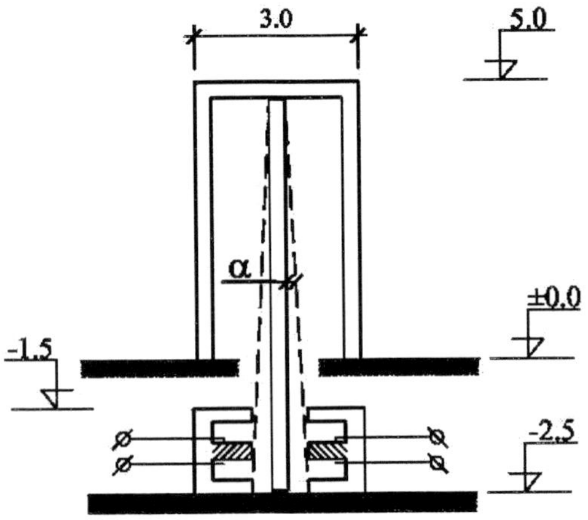

Figure 3.7. Frame diagram with Moon beam.

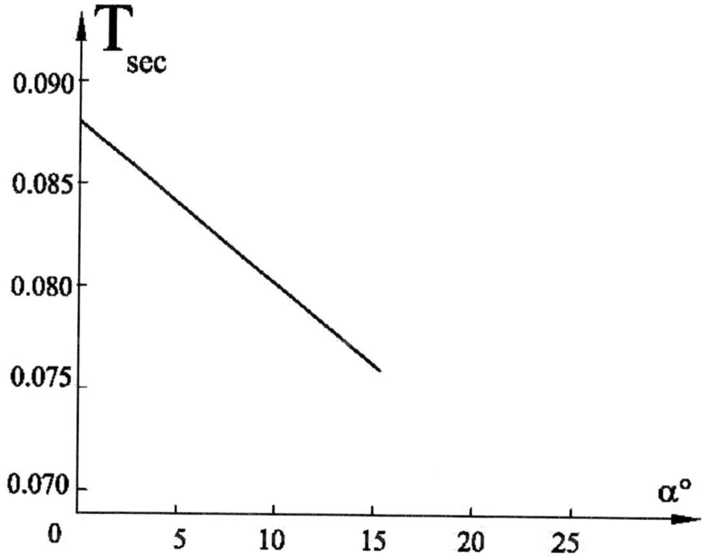

Figure 3.8. The relation of frame natural vibration period to Moon beam deviation angle.

Frame span L=3.0 m; frame height H=5.0 m; frame columns and collar-beams with pipe section diameter - 140 mm; wall thickness - 40 mm; Moon suspended beam of strip steel is of 100 mm width and 20 mm thickness; cross-section area A=20 cm^2.

Cantilever beam linear mass is:

$$m=\frac{q}{g}=\frac{10\cdot 2\cdot 7,81}{9,81}=0,167 \text{ grsec}^2\text{/cm}=0,000167 \text{ kgsec}^2\text{/cm}$$

E =210000 kg/cm^2 is a design resistance in stretching and bending plane:

$$I=\frac{bh^3}{12}=\frac{10\cdot 2^3}{12}=6,67cm^4$$

$$S_\delta = 1,5X1,5=0,75m$$

Insert numerical data:

$$\overline{I}=\sqrt{\frac{2\cdot 10000\cdot 0,5^2\cdot 10^7}{4\cdot 3,14\cdot 2353^2\cdot 0,75}}=30,9Amp\approx 31Amp$$

Wire area:

$$S=\frac{31}{2}=15,5mm^2$$

Wire diameter:

$$d_m^2 = \frac{4 \cdot 15,5}{3,14} = 19,74; \ \ \mathrm{d}_m = 4,44mm$$

Circular core diameter or rectangular sides perimeter of beam is:

$$a = 0,5X4 = 2,0 = 2000mm .$$

Perimeter of circular core coil:

$$a = 2\pi R = 6,28 \cdot 0,25 = 1,57m = 157mm .$$

Better take rectangle sides as a = b = 0.5 m.

$$\sum l = 2000 \cdot 2353 = 4,706 \cdot 10^6 mm .$$

Conductor resistance by formula is:

$$R = \rho \frac{\sum l}{S} = \frac{1,7 \cdot 10^{-5} \cdot 10^6}{15,5} = 5,16ohm .$$

Here $\rho = \frac{1}{\gamma}$ is resistivity of a conductor;
γ is electrical conductance of material;
S is conductor cross-section.
Moon beam natural vibrations frequency in magnetic field is defined with formula:

$$\omega^2 = \frac{4k\Phi_0^2}{a^3 m} - \frac{12,52EI}{ml^4} \tag{3.28}$$

where

$$\Phi_0 = \frac{\overline{I}N}{a} = \frac{31 \cdot 2353}{a} = 729,43 .$$

Here $\overline{I}$ is current force equal to 31 A;
N is number of coils equal to 2353;
α - the distance from the conductor to field point is equal to 100 cm.
Then

$$\omega^2 = \frac{4 \cdot 1 \cdot 532068 = 12}{100^3 \cdot 0,00016} - \frac{14 \cdot 10^6 \cdot 12 = 52}{750^4 \cdot 0,00016} = 13298,24$$

ω=115.3 rad/sec; ω= 18.36 Hz.

Moon beam nonlinearity coefficient is:

$$\gamma = 6,91 \frac{EF}{ml^4} = 6,91 \frac{2100000 \cdot 20}{0,00016 \cdot 750^4} = 5,73$$

Then Moon beam differential equation of Duffing type will be:

$$\frac{d^2T(t)}{dt^2} - 18,36^2 T(t) + 5,73 T^3(t) = 0$$

Stiffness in case of bending EI=2100000·6,67=14·10^6kgcm^2;
Stiffness in case of stretch EA=2100000·20=42·10^6kg;
Length of suspended beam L=750cm=7,5m.

Moon beam Duffing equation was solved according to the following rounded off initial data in program “MATCAD Professional”:

$$\frac{d^2}{dt^2} x(t) + c \frac{d}{dt} x(t) - wx(t) + rx^3(t) = A \cdot \cos(\omega t) \qquad (3.29)$$

where c =0.5

$\omega = 0$

A = 0

r=6.0

w=18.0

Phase picture is received as a result of Duffing equation solution (Figure 3.9). Figure 3.8 also gives the relation of frame natural vibration period to Moon bean angle deviation.

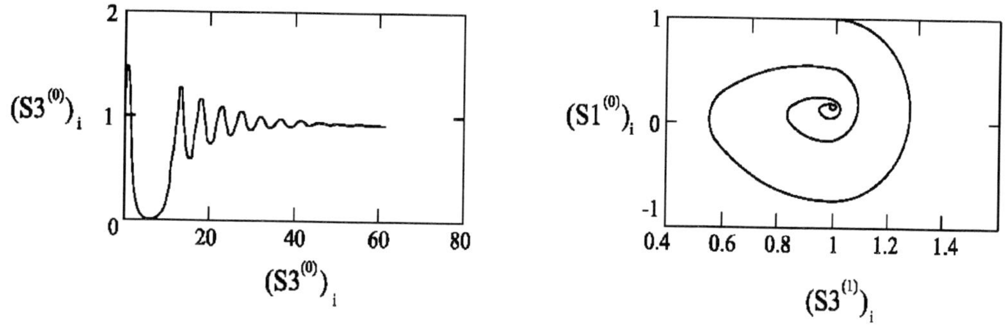

Figure 3.9. Phase picture of vibrations of the frame with Moon beam.

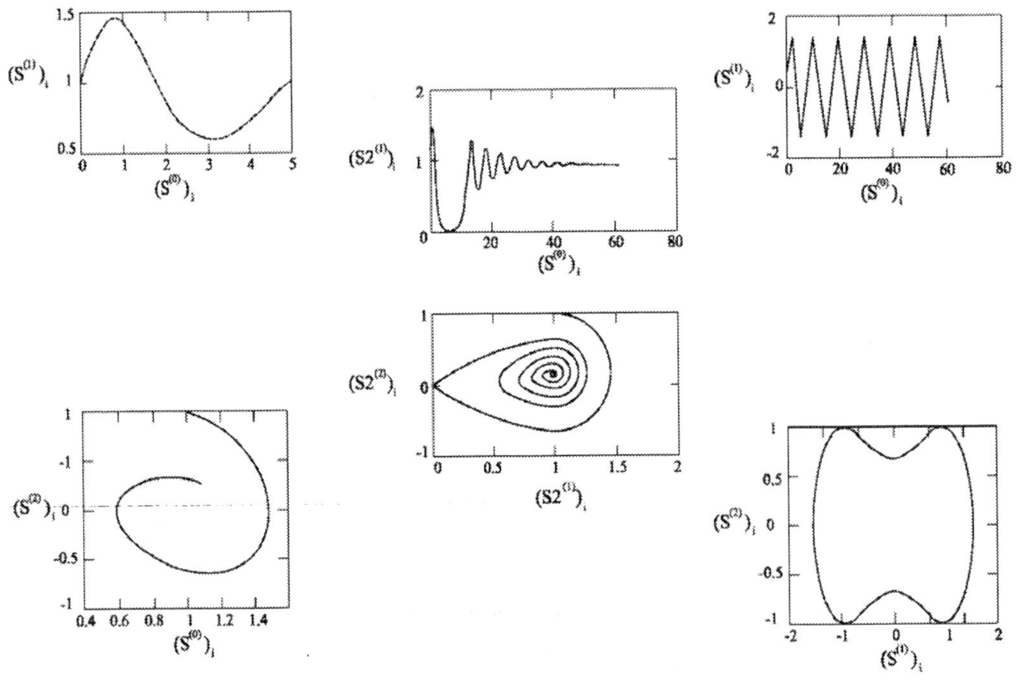

Figure 3.10. Phase pictures of vibrations of the frame with Moon beam.

As a conclusion we can note that using of Moon beam for vibrations damping is quite possible in particular conditions. Natural vibration period for the frame changes by linear law according to Moon beam deviation angle. According to phase trajectory the maximum rate of vibrations is:

$$V_{max}=Y_{max}\cdot\omega=1{,}464\cdot 18=26{,}3 \text{ cm/sec}$$

This problem is solved with MATCAD program. The solution is given in Figure 3.10.

3.4. Vibration Damping in Frames with Combined Collar-beams

Let's consider the behaviour of a frame with combined collar-beam affected upon with dynamic forces of seismics and wind.

Frame with combined collar-beam consists of posts rigidly connected to collar-beam and anchored in foundation.

Combined collar-beam is presented in two versions with parallel tie-bar and strut.

It must be mensioned that parallel tie bar, as well as strut can be prestressed or stressless (Figure 3.11.).

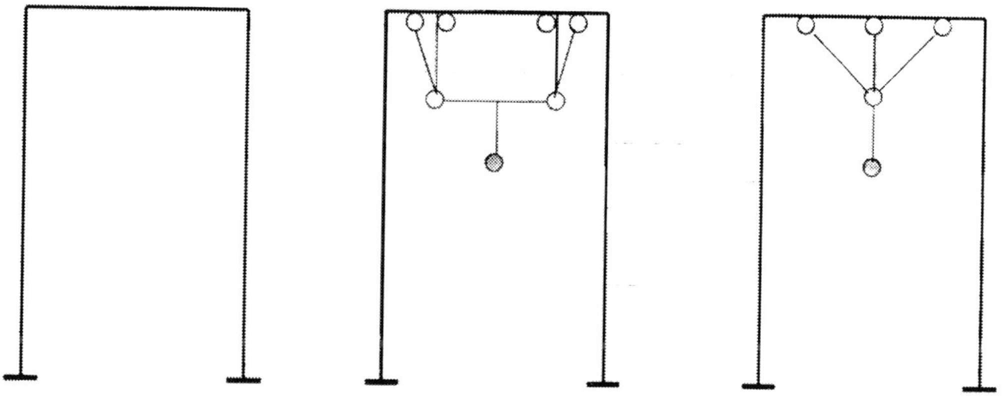

Figure 3.11. Frame diagrams with combined cross-bars:
a) simple frame, b) frame with combined collar-beam (tie-bar and pendulum), c) frame with combined collar-beam (strut and pendulum).

Stressless version is considered.

In both cases, parallel tie-bar, as well as, strut have load suspended in the middle, though suspension point may be in any place.

The suspended load can be considered as physical or mathematical pendulum according to the diameter of suspension tie-bar and the dimensions of load.

The behaviour of mathematical and physical pendulum according to immovable suspension point have been studied thoroughly enough.

In frame system operation the maximum amplitude and pendulum vibration state are of particular interest as with consideration of nonlinearity when pendulum deviation angles are different and pendulum suspension length changes, the vibration deviation and motion stability are changed respectively which is connected with pendulum load impacts to frame posts and failure of vibration isochronism.

In order to define motion state and maximum amplitude the numerical test has been carried out with program "MATCAD-2000".

Pendulum motion equation was considered with unit frequency during 5 seconds.

When considering the analogous problem it has been noted that accuracy degradation for angles approximating $180^0(\pi)$ is regarded as critical value: at transition to this value the character of motion changes - instead of vibrations we get rotation.

The same picture is noted when investigating the equation:

$$\frac{d^2}{dt^2}\theta(t) \mp \sin\left[\theta(t)\right] = 0 \qquad (3.30)$$

with the only difference that accuracy degradation, when determining the vibration amplitudes, began when angle of deflection got over $81^0\,48'$.

The example of a boundary value problem is given in Figure 3.12.

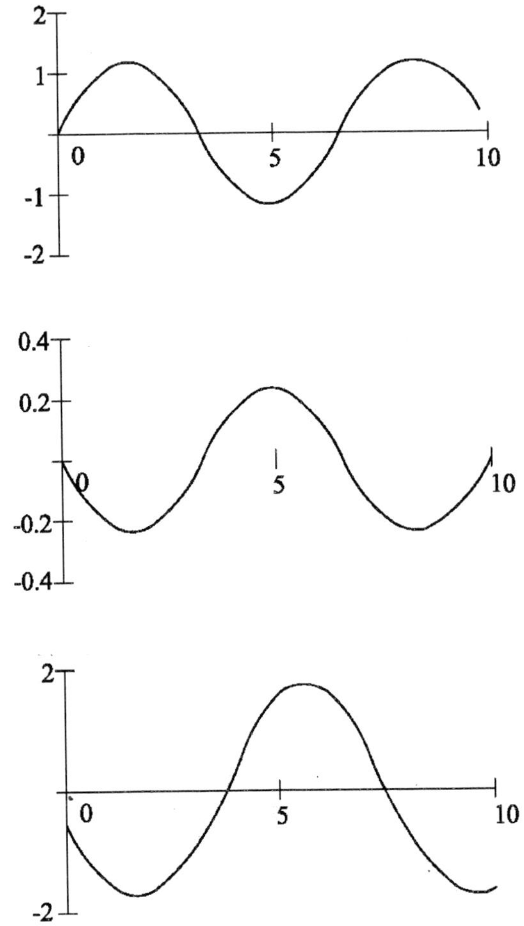

Figure 3.12. Pendulum vibration at different initial angles of deflection.

At different initial angle of deviation the similar length pendulums vibrate with different amplitudes but the periods of their vibrations will be similar if pendulum deviation angle does not exceed several degrees (1÷3)

The amplitude of vibrations of one pendulum will be more, that of another will be less, but amplitude duration for both pendulums will be similar. This is the peculiarity of the phenomenon of pendulums isochronic vibration. But as it is seen from numerical experiment this phenomenon at large deflection of angles is infringed and this is to be considered at frame system operation when suspended pipelines or suspended pendulums are used as frame vibrations dampers.

In this case the point of suspended pendulum anchoring is not fixed and makes periodical horizontal and vertical movements.

For the analysis of joint operation of parallel tie-bars and suspended pendulum under dynamic action the separated or independent method can be used.

Differential equation of parallel tie-bar motion in air flow can be expressed with Duffing equation:

$$m\ddot{y} + a_1 y + a_3 y^3 = P(t) \tag{3.31}$$

where P(t) is Carman's exciting force:

$P(t)=0{,}5\rho v^2 C_R S \sin\theta t$

$a_1=4H/l$; $a_3=8EF/l^3$

Here H is a bunton in tie-bar;

l is brace length;

EF is brace rigidity;

ρ is air density;

v is air flow rate;

C_R is total aerodynamic coefficient:

$\theta = Sh \cdot v / d$

Sh is Strauhal number;

d is brace cross-section.

The amplitude of induced vibrations of the brace is determined from the following expression:

$$A^3 + A\left(\bar{\omega}^2 - \theta^2\right)\frac{3}{4h} - \frac{2}{3h}\rho V^2 C_R S = 0 \tag{3.32}$$

where $h=8EF/Ml^3$

M is brace mass.

Lateral oscillations of brace change the location of gravity center of physical pendulum (Figure 3.13).

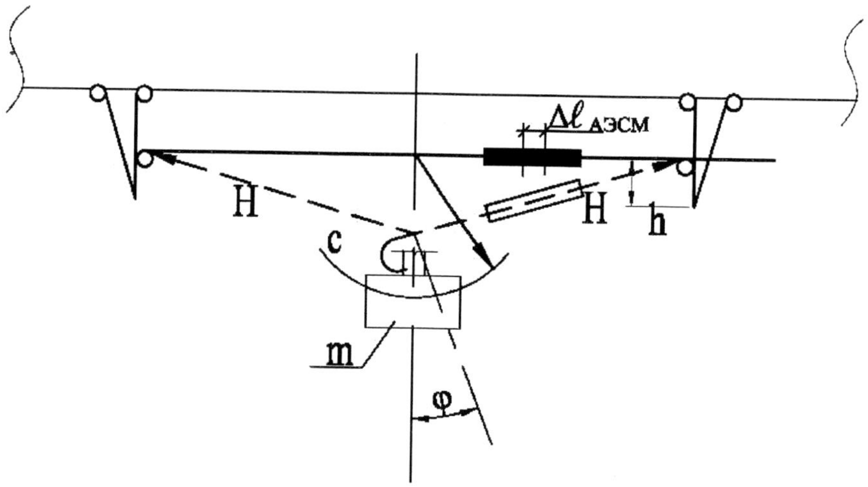

Figure 3.13. Diagram of parallel brace (brace and pendulum).

Kinetic and potential energy of physical pendulum will be:

$T=0{,}5I\varphi^2$ $v=v_1+v_2$

$$v_1 = 0{,}5C\frac{r}{R}\varphi^2; \quad v_2 = m(R+y)(1-\cos\varphi)g$$

where

I is inertia moment of physical pendulum;

C is the stifeness of physical pendulum suspension;

g is free fall acceleration;

r is the height of suspension;

R is gravity center height of physical pendulum;

y is ordinate.

The second order Lagrange equation has the following form:

$$\frac{d}{dt}\left(\frac{\partial T}{\partial \dot{q}_r}\right) - \frac{\partial T}{\partial q_r} + \frac{\partial v}{\partial q_r} = Q_r \tag{3.33}$$

If suppose that $Q_r=0$, we get system motion equation with periodical coefficients:

$$\frac{\partial^2\varphi}{\partial t^2} + \omega_0^2\left(1-\mu\cos\overline{\omega}t\right)\varphi = 0 \tag{3.34}$$

where

$$\omega_0 = \frac{cr}{IR} + \frac{mgh}{I}$$

$$\overline{\omega}^2 = \omega^2 - \frac{3}{4}hA^2;$$

$$\mu = mgA / I\omega_0^2$$

Here A is amplitude which is received as a result of solution of brace equation.

As it is known the first area of instability for Mathieu equation is determied from inequality:

$$1-\frac{1}{2}\mu+\cdots \le \frac{4\omega_0^2}{\overline{\omega}^2} \le 1+\frac{1}{2}\mu\cdots \tag{3.35}$$

The effect of a cross-bar combined of brace and suspended pendulum on the operation of frame system under dynamic action of wind and seismics is consider on example of single span, one storey steel frame with the following initial data:

Frame span - L=3,0m;
Frame height – H=5,0m
Frame beam and columns made of pipes with section diameter – 140X140mm;
Brace with cable diameter – 12,0mm;
Pendulum suspension with cable diameter – 12.0 mm;
Pendulum weight: G=2400 kg.

Brace anchors are located in the distance of 0.5 m from frame joint. Seismic action is 8 points of MSK-64.

The frequency of frame natural vibrations in the first form for simple frame is 5.995 Hz (period 0.167 sec). The joint point of cross-bar and column has displaced horizontally for 0.004467 m.

The variation of frame joint displacement because of load in horizontal direction for simple systems is of linear dependence.

The frequency of natural vibrations of a frame with strut and pendulum made, by the first form, 5.932 Hz and maximum amplitude, by the first form, is 0.004586 m. For that with parallel braces maximum amplitude is 0.001838 m.

Calculation was performed using programs "LIRA" and "MIRAZH".

As a result of carried out numerical modeling (Figure 3.14) for investigation of frame dynamic behavior we conclude that using brace and suspended pendulum, dynamic characteristics in frame systems, particularly frequency (period) and vibration amplitude can be significantly varied.

The best result is received when vibration amplitude decreases 2.4 times. At the same time nonlinear behavior of a pendulum (magnitude of deflection angle), as well as united vibrations of brace and pendulum and their vibrations within stability areas effect the regulation of dynamic characters in frame system.

№ Load	№ Mode	Eigenvalue	Frequency		Period
			rad/sec	1/sec	
2	1	0,027	0,027	37,666	0,167
2	2	0,016	0,016	62,569	0,100
2	3	0,009	0,009	111,939	0,056

Figure 3.14. The results of numerical modeling.

3.5. Frame Systems with Diagonal Bracing and Dampers Placed on them

In the practice of railway electrification rigid cross-pieces (cross-bar anchoring supports) are used for getting anchor forces of different contact wires.

The supports of power lines also have the frame structure form with suspension insulator strings of power lines.

Frame supports are used for suspension of pipelines in one and two storey single span frames.

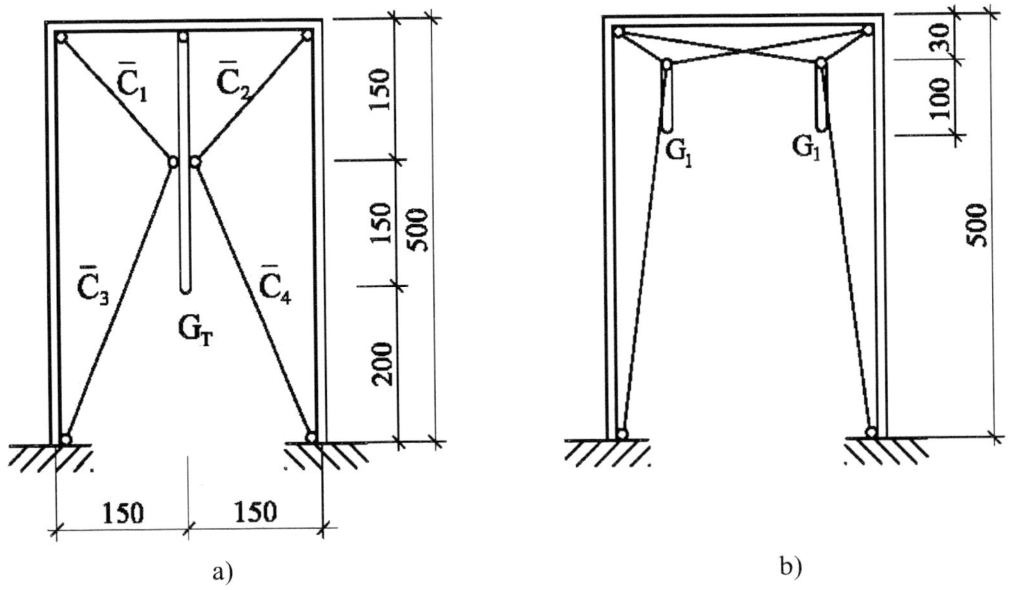

Figure 3.15. Diagram of a frame with symmetrical diagonal bracings and pendulums.

The span structure of suspended bridges is also suspended on portal frames with pendulum type vibration dampers installed in constructions. In the supports of power lines and suspended pipe lines besides П-shaped frames there are used the A-shaped frames as well.

Frame system with diagonal bracings and pendulum is a complex system consisting of two separate systems: frame construction itself and diagonal bracings connected to each other with a pendulum.

Here, the word "bracing" means that the vibrations of one system affect another system and vice versa.

For physical analysis of the phenomenon in a complex system it is necessary to know the nature of vibrations in separate "partial" systems which make the complex system.

Partial system is received from a complete system when we have "rigid" anchoring of all joints except the given one.

In the considered case such limitation was done to the frame. Diagonal bracings with pendulum vibrate in drawing plane, as well as horizontally to drawing plane.

Consider two frame systems with diagonal bracings and pendulum: the first frame with symmetrical diagonal bracings and suspended pendulum in the middle of frame beam. The second with pendulum suspended in equal distances from upper joints of frame (Figure 3.15a, b).

Frame systems can vibrate longitudinally, as well as laterally to frame plane.

3.5.1. Motion Equation of the Pendulum with Diagonal Bracings

The diagram of diagonal bracings with pendulum is given in Figures 3.16 and 3.17.

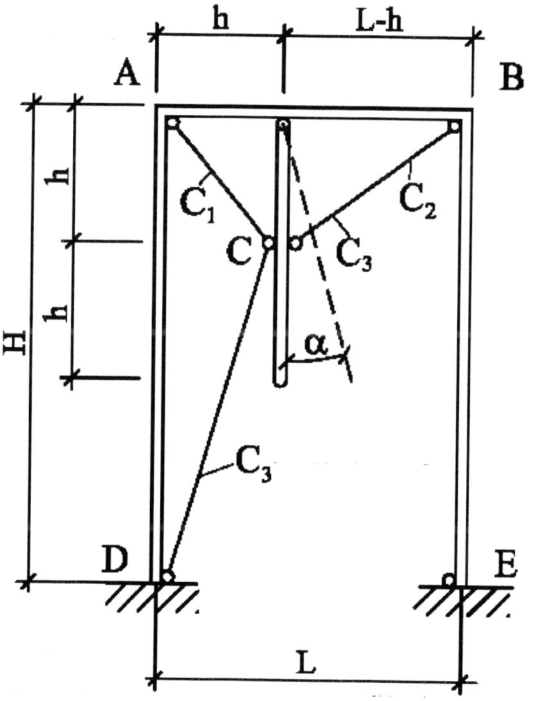

Figure 3.16. General view of a frame with asymmetric diagonal bracings and pendulum.

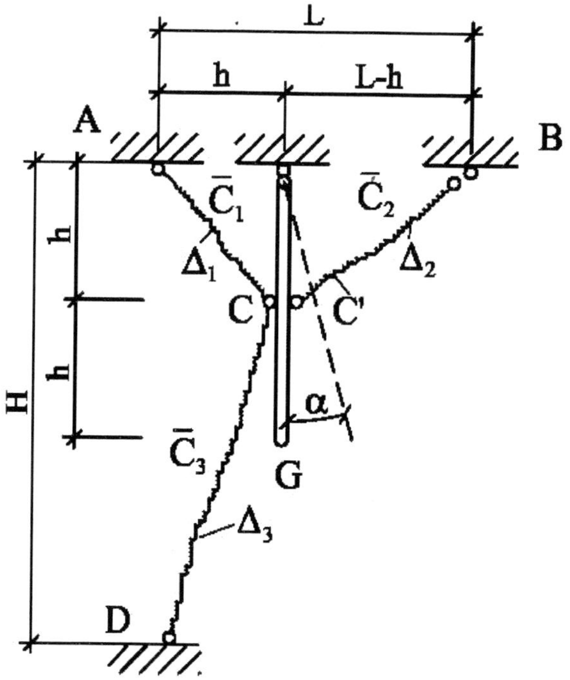

Figure 3.17. Design scheme of diagonal bracings and pendulum.

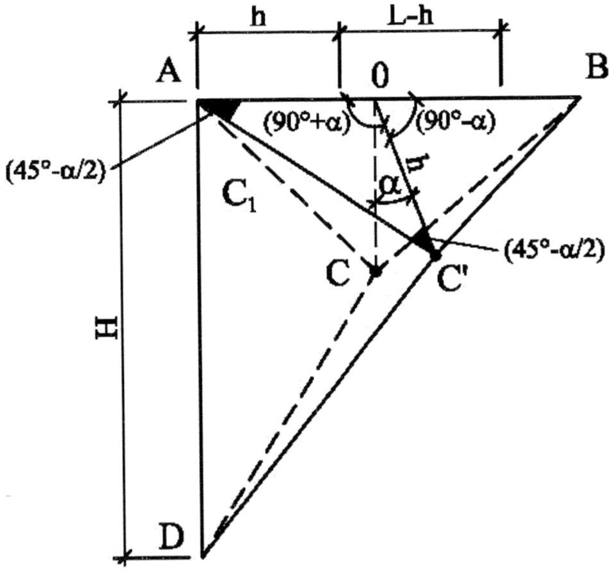

Figure 3.18. Geometrical diagram of bracings.

Potential energy of the system is:

$$\sum \ddot{I} = \ddot{I}_1 + \ddot{I}_2 + \ddot{I}_3 + \ddot{I}_{iayo} = \frac{\overline{N}_1 \Delta_1^2}{2} + \frac{\overline{N}_2 \Delta_1^2}{2} + \frac{\overline{N}_3 \Delta_1^2}{2} + G\Delta h \qquad (3.36)$$

$$\ddot{I}_{iayo} = G\Delta h = G\left(h - h\cos\alpha\right) \qquad (3.37)$$

Kinetic energy of the system is:

$$\dot{O} = \frac{1}{2}\frac{G}{g}h^2 \cdot \alpha^2 + \frac{1}{2}\cdot\frac{1}{3}\frac{G}{g}h^2 \cdot \alpha^2 \qquad (3.38)$$

Use the theorem of cosines (Figure 3.18):

$$\left(AC'\right) = h^2 + h^2 - 2h \cdot h\cos\left(90^\circ + \alpha\right)$$

$$\left(BC'\right) = h^2 + \left(L - h\right)^2 - 2h \cdot \left(L - h\right)\cos\left(90^\circ - \alpha\right)$$

$$\left(DC'\right) = H^2 + \left(AC'\right)^2 - 2H\left(AC'\right)\cos\left(45^\circ + \frac{\alpha}{2}\right)$$

The increment of bracing lengths is:

$$\Delta_1 = AC' - AC = \sqrt{h^2 + h^2 - 2h \cdot h\cos(90^\circ + \alpha)} - \sqrt{2} \cdot h$$

$$\Delta_2 = BC - BC' = \sqrt{(L-h)^2 + h^2} - \sqrt{h^2 + (L-h)^2 - 2h(L-h)\cos(90^\circ - \alpha)}$$

$$\Delta_3 = DC' - DC = \sqrt{H^2 + (AC')^2 - 2H^{(AC')} \cdot \cos\left(45^\circ + \frac{\alpha}{2}\right)} - \sqrt{(H-h)^2 + h^2}$$

Second degree Lagrange equation:

$$\frac{d}{dt}\left(\frac{\partial T}{\partial \dot{q}}\right) - \frac{\partial T}{\partial q} + \frac{\partial \ddot{I}}{\partial q} = Q$$

for natural oscillations: Q=0, here $\frac{\partial T}{\partial q} = 0$.

Determination of derivatives of potential and kinetic energy:

$$\ddot{I}_{iayo} = G\Delta h = G(h - h\cos\alpha)$$

$$\frac{\partial \ddot{I}_{iayo}}{\partial \alpha} = Gh\sin\alpha$$

$$T = \frac{2}{3}\frac{G}{g}h^2 \cdot \alpha^2; \quad \frac{\partial T}{\partial \alpha} = \frac{4}{3}\frac{G}{g}h^2\left(\frac{d^2\alpha}{dt^2}\right)$$

$$\frac{\partial \ddot{I}_{iayo}}{\partial \alpha} = \frac{\partial}{\partial \alpha}\left(\frac{\overline{N}_1\Delta_1^2}{2} + \frac{\overline{N}_2\Delta_1^2}{2} + \frac{\overline{N}_3\Delta_1^2}{2}\right) + G\Delta h\sin\alpha$$

The equation of motion of partial system with damping taken into consideration, has the following form:

$$\frac{d^2\alpha}{dt^2} + \varepsilon\frac{d\alpha}{dt} + \omega^2\alpha + \xi\alpha^2 + \beta\alpha^3 + \gamma\alpha^5 + P = 0 \qquad (3.39)$$

where ω^2 is the frequency of natural oscillations of the partial system;

ξ, β, γ are nonlinearity coefficients of the system;

P is the free member;

ε is the coefficient of damping.

The frequency of natural oscillations is determined with formula:

$$\omega^2 = a_1 + a_2 + a_3 + a_4 + a_5 = \sum_{i=1}^{i=5} a_i \qquad (3.40)$$

where

$$a_1 = \frac{3g}{4h}$$

$$a_2 = \frac{3}{8}\frac{g\overline{C}_1}{G}$$

$$a_3 = \frac{3g}{4Gh^2}\frac{\overline{C}_2}{2}\frac{16h^4(L-h)}{\left(4h^2+L^2-2Lh\right)^2}$$

$$a_4 = \frac{3g}{4Gh^2}\frac{\overline{C}_3}{2}\left(\frac{h^2}{2}-\frac{3Hh}{4}\right)\left[1-\frac{\left(H^2+3h^2-2Hh\right)}{2h\left(H^2+2h^2+2Hh\right)^{1/2}}\right]$$

$$a_5 = \frac{3g}{4Gh^2}\frac{\overline{C}_3}{2}\left[\frac{8h^4}{4\left(H^2+2h^2+2Hh\right)}\right]$$

In particular case when $\overline{C}_1 = \overline{C}_2 = C$ and $\overline{C}_3 = \overline{C}_4 = 0$ the frequency of natural oscillations of the frame with bracings and pendulum equals:

$$\omega = \sqrt{\frac{3g}{4}\left(\frac{C}{G}+\frac{1}{h}\right)} \tag{3.41}$$

which coincides with the presented expression.

3.5.2. Numerical Example'

For example consider two diagrams of the frame with diagonal bracings and pendulums (see Figure 3.15 a,b).

Initial geometrical data are given in Figure 3.15 a,b for diagram. The frame is made of pipes with diameter D = 140 mm and wall thickness t = 36 mm, diagonal bracings are plain reinforcement with diameter D = 12 mm, suspended pendulum is of round steel with diameter D = 32 mm, load at the end G = 0 12500 kg. The coefficient of stiffness of diagonal bracings is:

$$\overline{C}_1 = \overline{C}_2 = 10670 kg/sm;\ \overline{C}_3 = \overline{C}_4 = 5953 kg/cm$$

For the diagram given in Figure 3.15b the frame is made of pipes with diameter D = 140 mm and wall thickness t = 40 mm. Diagonal bracings and pendulum suspension are made of cable with diameter D = 12.0 mm.

On the end of suspended pendulum the loads, each weighting G = 2400 kg, are located.

The calculation of the first diagram was done in program SAP 2000. The vibrations of frame system happened vertically to frame plane.

The first period of natural vibrations of frame system in the first mode was T = 0.4456 sec.

Without diagonal bracings and pendulum and with the load in the middle of frame cross-piece the oscillation period made T = 0.4410 sec.

Hence, the existence of diagonal bracings with suspended pendulum changes the period of frame system natural oscillations for 7.4 %.

The first mode of frame system vibrations is given in Figure

The calculation of the second diagram is performed in program "LIRA-8.2".

The comparison of vibration amplitudes (in drawing plane) to a simple frame and a frame with diagonal bracings and two pendulums showed that in the first case the deviation was 4.467 cm and in the second case – 0.148 cm.

For simple frame the vibration period in the first mode is T = 0.167 sec. For the frame with diagonal bracings and pendulum T = 0.7 sec. The difference made 76 % (Figure).

Thus, it can be concluded that for damping of vibrations in frames the diagonal bracings with suspended loads (pendulum type) can be successfully used which significantly changes the frequency of vibrations (period) and decreases amplitudes.

3.6. Experimental Research of Steel Frames with Combined Dampers

In order to continue earlier carried out test researches for static and dynamic effects the steel frame have been experimentally investigated. Frame operation was studied in two ways: by physical and computer modeling, which have two types of dampers: combined dampers with horseshoe-shaped devices and round link dampers.

Test model (Figure 3.19, 3.20, 3.21, 3.22, 3.23) represents a closed type frame made of tubular members with D = 34 mm, pipe wall thickness t = 3.5 mm, The dimensions of frame model are: span l = 55.0 cm; frame height h = 122.0 cm; distance from frame support to its first collar-beam - 9.0 cm. Inside the frame a damper with two types of bracings is located.

The first type of a damper is a horseshoe circle with diameter - 15.3 cm, thickness - 6.0 mm and width 2.5 cm. The clearance in the circle is 3.0 cm.

Bracings are made of cables of 1.8 mm diameter.

The horseshoe circle is fixed in the middle of collar-beam with rubber pod using shock absorber. A trussed construction is installed in 16.0 cm from the upper collar-beam of the frame.

The second type of a damper represents a closed circle with diameter - 15.3 cm, thickness - 6.0 mm and width 2.5.cm. The circle is installed in the middle of frame span on the half height of the frame. In four points of the circle bracings of diameter 1.8 mm are fixed in diagonal direction.

On the upper collar-beam of frame model an electric motor with load is mounted. In the upper joint of the frame static load P = 120-140 decaN (1 decaN – decaneuton = 1 kgf) is applied with the help of horizontal cable and with the use of dynamometer.

Figure 3.19. View of experimental model of the frame with ring bracing.

Figure 3.20. The used measuring device.

Figure 3.21. View of experimental model of the frame with horseshoe-shaped bracing.

Figure 3.22. The used measuring device.

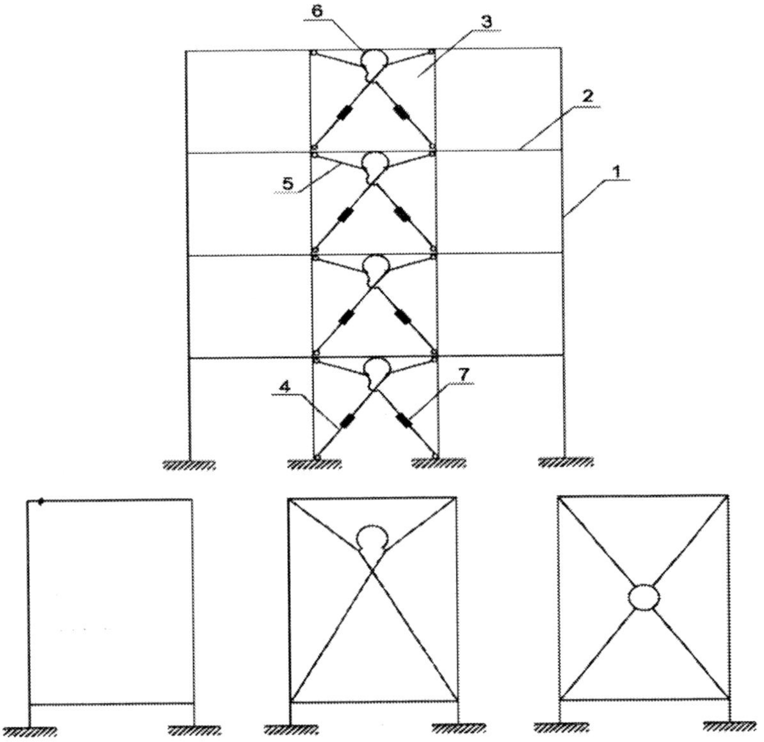

a

b

Figure 3.23. Diagrams of frame test models with combined dampers of vibrations.

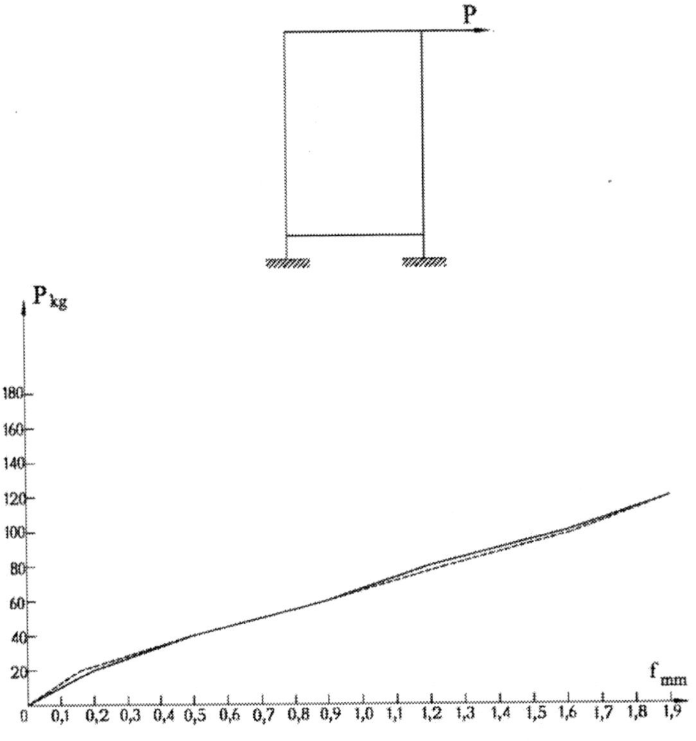

a)

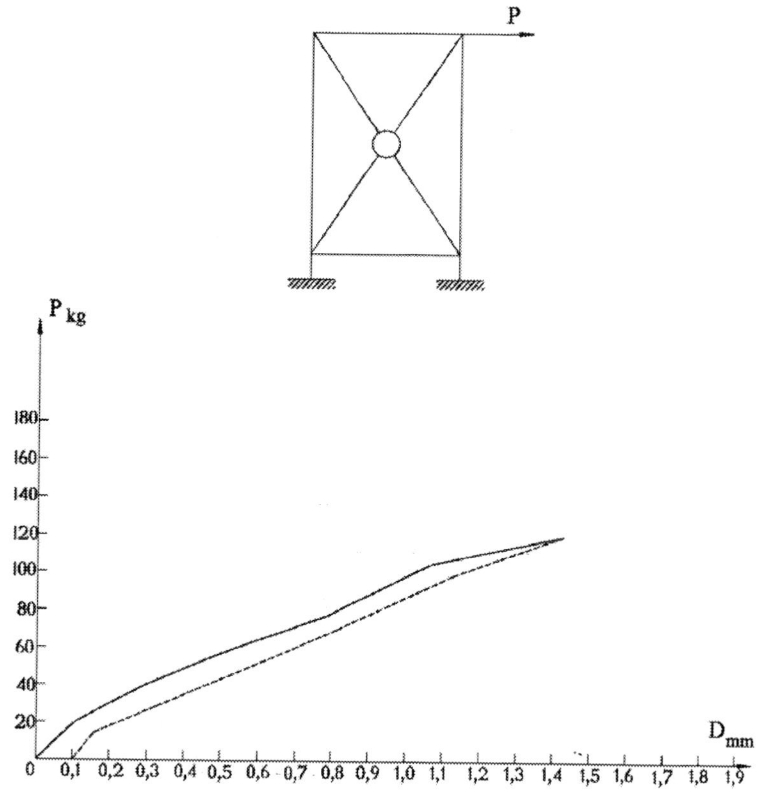

b)

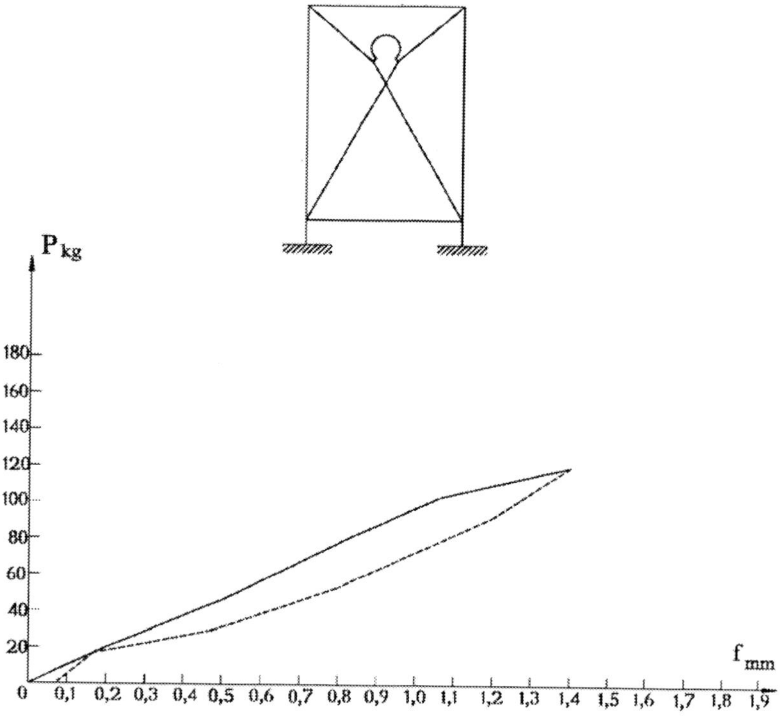

c)

Figure 3.24. Graph of frame displacement: ——— loading - - - - -unloading.

Table 3.2.

Frame displacement caused by static load (Figure 3.24 a,b,c, d,e)		Coefficient of frame oscillation amplitude reduction caused by dynamic load	
Frame designation	Δ mm	Frame designation	K
Simple frame	1.9	Simple frame	1.0
Frame with circular damper	1.35	Frame with circular damper	1.14
Frame with horseshoe damper	1.38	Frame with horseshoe damper	1.18

Static loading in joint was transmitted step-by-step with 20 decaN (six steps in all) and horizontal displacement of frame was measured with Macsimov's deflection meter. Readings are given in the form of a table.

Dynamic loading of frame was done with electric motor with eccentrically arranged load. The registration of vibrations was done with oscillograph HO441 and vibro-sensor KH001Г.

The number of electric motor revolutions was registered with tachometer ИО-30.

Vibrations were recorded on paper film with time mark of 0.1 sec.

Vibrations were recorded under static load applied in the upper joint of the frame P = 60 decaN and in case of bracings tension equal to 50 decaN.

The frequency of forced oscillations made θ=18 sec. The results of processing of oscillograph records are given in the Table.

Computer modeling of frame with combined damper is accomplished in programs SAP-2000" and "LIRA-8".

While using the both programs the horizontal load P = 150 daN applied in the upper joint of the frame is considered. The horizontal cable as well as horizontal bracing of upper joint is not taken into account.

Therefore, the values of frame system periods and dynamic displacements differ from experimental data and only their qualitative part is preserved and enables to estimate the effect using.

As a result of tests and computer modeling it can be concluded that using of two types of dampers in frames significantly change vibration periods and decrease dynamic displacements.

Compared to the usual frame the coefficient of amplitude decrease without damper is K = 1.14 for circular damper and K = 1.18 for horseshoe damper, i.e. frame oscillation amplitude decreases for 18 % (Table 3.2).

3.7. Regulation of Strains of Prestressed Beams

Generally collar-beams are fixed to frame columns jointly and rigidly (Figure 3.25). Joint support of the collar-beam on frame columns enables to consider it as the girder supported on two supports with prestressed parallel tie-bar on which АЭСМ is mounted (Figure 3.26, 3.27).

Double-T beam loaded with static load is taken as prestressed steel girder.

The displacement of not preliminarily tensioned beam as a result of uniformly distributed and concentrated loads is expressed with formula:

$$f = \frac{5}{384}\frac{q^{\sigma}\ell^4}{E\Im} + \frac{P^{\sigma}\ell^3}{48E\Im} \quad (3.42)$$

Here q^{σ} is equdistributed standard load per one meter of girder;

P^{σ} is concentrated standard load;

ℓ beam span;

$\Im$ inertia moment of the girder;

E is module of flexibility.

Double-T beam is I N 16 (cross-section area A = 20.0 cm^2, $\Im$=873cm^4; q^{σ}=15daN/m; R_y=2100daN/cm^2; E=2100000 daN/cm^2).

1daN = 10 N = 1 kg is taken as dimension.

Freely supported beam span l = 324 cm.

The displacement of prestressed beam because of static load can be expressed with the following approximated formula:

$$f = \frac{1}{48E\Im}\left(p^{\sigma}\ell^3 + \frac{5}{8}q^{\sigma}\ell^4 - 6\left(X_1 + X_2\right)\bar{h}\ell^2\right) \quad (3.43)$$

Here X_1 and X_2 are prestressed and selfstrain forces;

$\bar{h}$ is the distance from guy-rope axis to neutral axis.

Test model of steel beam represented a double-T beam I N16, steel =st.3; parallel tie-bar with diameter Ø2.1 mm, length of cable-tie bar L = 1600 mm; beam span l = 3240 mm.

In the middle part of the beam on its upper shelf an electric engine was mounted.

Static displacements and vibration amplitudes were measured with Macsimov's deflection meter.

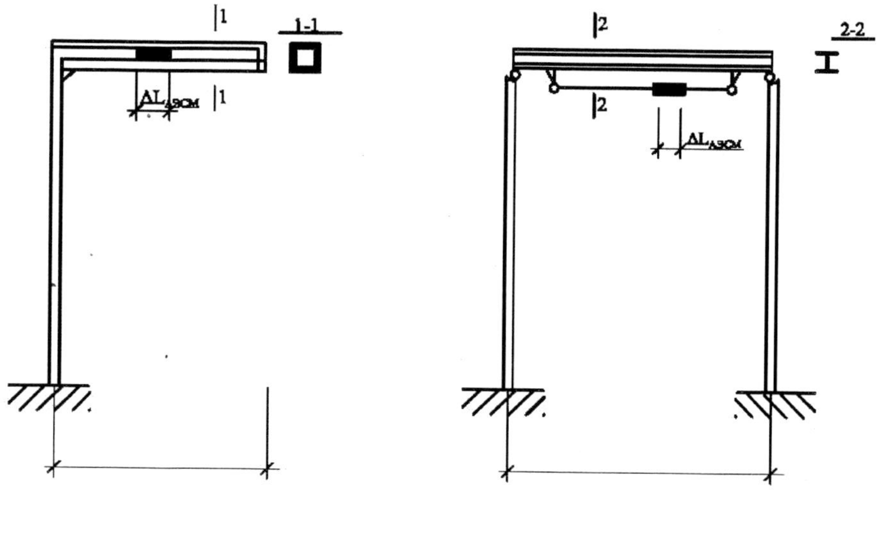

Figure 3.25. Diagrams of prestressed frames.

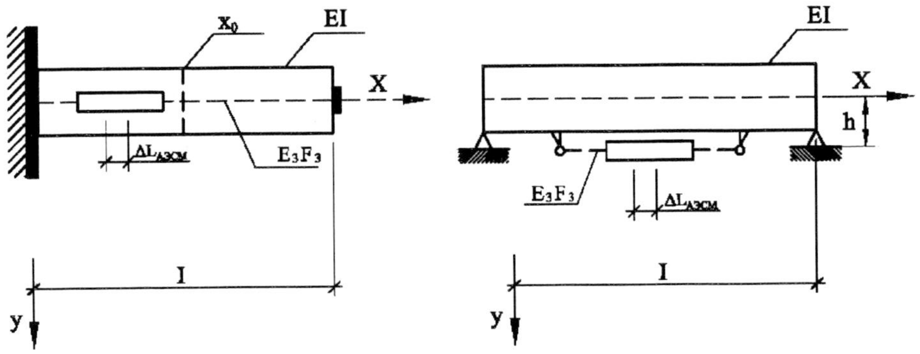

Figure 3.26. Diagrams of prestressed frames.

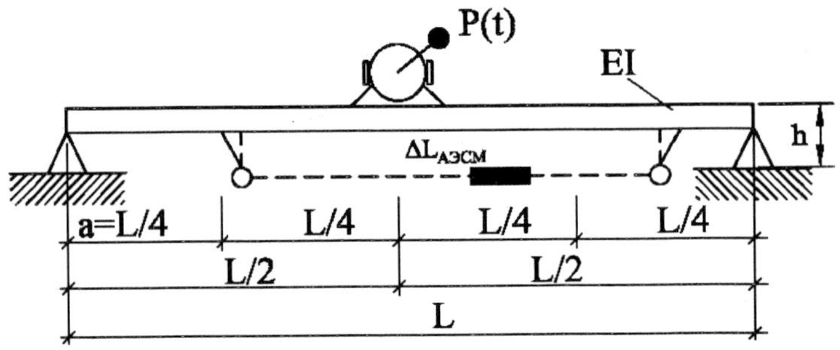

Figure 3.27. Diagrams of prestressed beam under dynamic load.

Concentrated load is applied step-wise in the middle part of the beam and when tension in brace and beam displacement were measured achieved 500 daN (5.0 kN).

When static concentrated load achieved P = 500 daN (5.0 kN) and tension in the tie-bar achieved S = 105 daN (1.05 kN) the displacement in the middle part of the beam made 0.9 cm, unstrained beam displacement made 1.2 cm. The decrease of displacement was 8.0 – 25% (Figure 3,28, 3.29, 3.30, 3.31, 3.32).

Here "MMA" means Macsimov's apparatus, "IND" – indicator-deflectometer

The exact expression for the curve of a beam or a column is:

$$\chi = \tag{3.44}$$

Hence we get

$$E\Im(x)\frac{d^4y}{dx^4} = q(x) \tag{3.45}$$

With concentrated forces taken into consideration:

$$E\Im(x)\frac{d^4y}{dx^4} = q(x) + \sum_{i=1}^{i=n} P_i\delta(x-a_i) + \sum_{i=1}^{i=n} M_i\delta'(x-b_i) + \cdots \tag{3.46}$$

where δ is Dirack function.

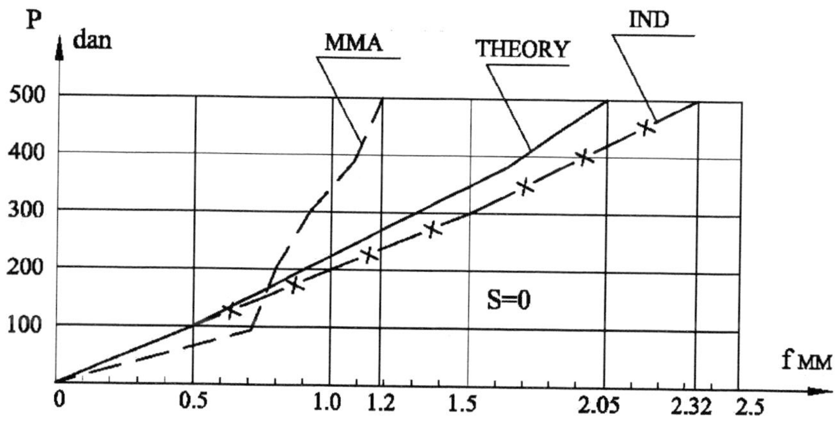

Figure 3.28. Graphs of theoretical and experimental results of deflections, S=0.

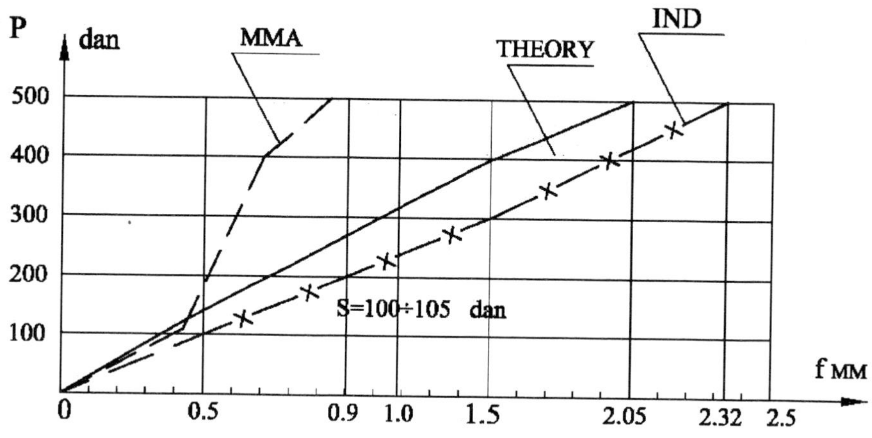

Figure 3.29. Graphs of theoretical and experimental results of deflections, S=100÷105daN.

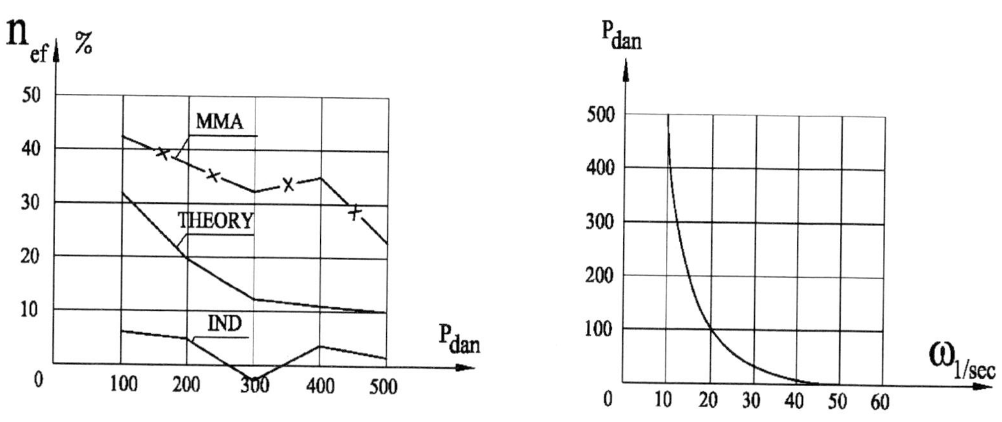

Figure 3.30. Efficiency of beam prestressing for deflection and natural vibration frequency.

In our case we shall have the following expression:

$$E\Im(x)\frac{d^4y}{dx^4}=q(x)+P(t)\delta(x-a)+M(t)\delta'(x-b)-M(t)\delta'[x-(\ell-b)] \quad (3.47)$$

Using d'Alambert principle from static equation we get dynamic equation:

$$E\Im(x)\frac{d^4y}{dx^4}=\overline{m}(x)\frac{\partial^2y}{\partial t^2}+\overline{M}\delta(x-a)\frac{\partial^2y}{\partial t^2}=P(t)\delta(x-a)+M(t)\delta'(x-b)-M(t)\delta'[x-(\ell-b)]$$

(3.48)

Time dependent factors:

$P(t)=P_0\cos\omega t;$ $M(t)=M_0\cos\omega t$

$y(x,t)=W(x)\cos\omega t;$ $y^{IV}=W^{IV}(x)\cos\omega t$

and $\ddot{y}=-\omega^2W(x)\cos\omega t$

If we substitute equations into 3.48, we get:

$$\begin{aligned}&\frac{d^4W(x)}{dx^4}\cos\omega t-\frac{m(x)}{E\Im(x)}\omega^2W(x)\cos\omega t-\frac{\overline{M}}{E\Im(x)}\delta(x-a)\omega^2W(x)\cos\omega t=\\&=P_0\delta(x-a)\cos\omega t+M_0\delta'(x-b)\cos\omega t-M_0\delta'[x-(\ell-b)]\cos\omega t\end{aligned} \quad (3.49)$$

Accept the solution of equation (3.49) as $W(x)=f\sin\frac{\pi x}{\ell}$ with boundary value conditions satisfied:

$$\begin{aligned}&x=0;\ \ W(x)=0\\&x=\ell;\ \ W(\ell)=0\end{aligned} \quad (3.50)$$

If we use Bubnov-Galerkin method and determine:

$$\frac{d^4W}{dx^4}=f\frac{\pi^4}{\ell^4}\sin\frac{\pi x}{\ell}$$

we get:

$$\int_0^\ell f\frac{\pi^4}{\ell^4}\sin\frac{\pi x}{\ell}\sin\frac{\pi x}{\ell}dx-\int_0^\ell\frac{\overline{m}(x)}{E\Im(x)}\omega^2 f\sin\frac{\pi x}{\ell}\sin\frac{\pi x}{\ell}dx-$$
$$-\int_0^\ell\frac{\overline{M}(x)}{E\Im(x)}\omega^2 f\delta(x-a)\sin\frac{\pi x}{\ell}\sin\frac{\pi x}{\ell}dx=\int_0^\ell\frac{P_0}{E\Im(x)}\delta(x-a)\sin\frac{\pi x}{\ell}dx+ \quad (3.51)$$
$$+\int_0^\ell\frac{M_0}{E\Im(x)}\delta'(x-a)\sin\frac{\pi x}{\ell}dx-\int_0^\ell\frac{M_0}{E\Im(x)}\delta'\left[x-(\ell-b)\right]\sin\frac{\pi x}{\ell}dx$$

After integration when $E\Im(x)=const$ and $\overline{m}(x)=const$, we get:

$$f\frac{\pi^4}{\ell^4}\frac{l}{2}-f\frac{\overline{m}}{E\Im}\omega^2\cdot\frac{\ell}{2}-f\frac{\overline{M}}{E\Im}\omega^2\cdot 1=\frac{P_0}{E\Im}\cdot 1-\frac{M_0}{E\Im}\frac{\pi}{\ell}\cos\frac{\pi b}{\ell}+\frac{M_0}{E\Im}\frac{\pi}{\ell}\cos\frac{\pi(\ell-b)}{\ell} \quad (3.52)$$

or

$$f\left(\frac{\pi^4}{2\ell^3}-\frac{\overline{m}\omega^2\ell}{2E\Im}-\frac{\overline{M}\omega^2}{E\Im}\right)=\frac{P_0}{E\Im}-\frac{M_0}{E\Im}\frac{\pi}{\ell}\frac{\sqrt{2}}{2}-\frac{M_0}{E\Im}\frac{\pi}{\ell}\frac{\sqrt{2}}{2} \quad (3.53)$$

Here $M_0=\frac{E_sF_s\overline{h}_s}{L_s}\Delta\ell_{АЭСМ}$;

s index belongs to brace;

E_sF_s is brace rigidity including АЭСМ.

and hence we have:

$$f=\frac{\dfrac{P_0}{E\Im}-\dfrac{\pi\sqrt{2}M_0}{E\Im\ell}}{\dfrac{\pi^4}{2l^3}-\dfrac{\overline{m}\omega^2\ell}{2E\Im}-\dfrac{\overline{M}\omega^2}{E\Im}} \quad (3.54)$$

After solving the equation of motion (eq. 3.48) we define the gap between contactors with AECM device:

$$\Delta\ell_{АЭСМ}=\left[2P_0\ell^3L-fL\left(\pi^4E\Im-\overline{m}\omega^2\ell^4-2\overline{M}\omega^2\ell^3\right)\right]/\,2{,}82\pi E_sF_s\overline{h}_s\cdot\ell^2 \quad (3.55)$$

where P_0 is concentrated force on tie;

f is the amplitude of tie vibration;

ℓ is tie span;

L is tie-bar length including АЭСМ;

$\overline{m}$ is uniformly distributed mass on the tie;

$\overline{M}$ is concentrated mass on tie;
E_sF_s is tie-bar rigidity by АЭСМ;
$\overline{h}_i$ is the distance from tie-bar axis to tie axis;
$\omega = \dfrac{2\pi}{T}$ is the frequency of natural vibrations of a beam which defined with formula:

$$\omega^2 = \frac{E\Im \cdot R_{1x}}{\overline{m}R_{5x} + \overline{M}R_{7x}} \tag{3.56}$$

where

$$R_{1x} = \int_0^\ell \frac{d^4W(x)}{dx^4} W(x)dx; \quad R_{5x} = \int_0^\ell W^2(x)dx;$$
$$R_{7x} = \int_0^\ell \delta(x-a)W^2(x)dx; \tag{3.57}$$

When $W(x) = \sin\dfrac{\pi x}{\ell}$, we get:

$$\omega^2 = \frac{\pi^4}{2\ell^3(0,5+\mu)} \cdot \frac{E\Im}{\overline{m}\ell}; \quad \mu = \frac{\overline{M}}{\overline{m}\ell}; \tag{3.58}$$

or

$$\omega = \sqrt{\frac{48,606}{(0,5+\mu)} \cdot \frac{E\Im}{\overline{m}\ell^4}} \tag{3.59}$$

With consideration of axial force:

$$\omega^2 = \frac{E\Im\pi^4 - N\pi^2\ell^2}{2(0,5+\mu)\overline{m}\ell} \tag{3.60}$$

By J.W.Rayleigh formula:

$$\omega = \sqrt{\frac{48,0}{(0,485+\mu)} \cdot \frac{E\Im}{\overline{m}\ell^4}} \tag{3.61}$$

By E.Sekhniashvili formula:

$$\omega = \sqrt{\frac{49,15}{(0,504+\mu)} \cdot \frac{E\Im}{\overline{m}\ell^4}} \qquad (3.62)$$

Without concentrated mass (3.58 and 3.59), we have:

$$\omega = \frac{9,86}{\ell^2}\sqrt{\frac{E\Im}{\overline{m}}}$$

By E.Sekhniashvili formula:

$$\omega = \frac{9,875}{\ell^2}\sqrt{\frac{E\Im}{\overline{m}}}$$

By J.W.Rayleigh formula:

$$\omega = \frac{9,948}{\ell^2}\sqrt{\frac{E\Im}{\overline{m}}}$$

From the reference-book:

$$\omega = \frac{9,8596}{\ell^2}\sqrt{\frac{E\Im}{\overline{m}}}$$

Dynamic displacements of the beam were experimentally fixed with Macsimov's deflectometer according to the opening of the sector visually; when AECM was switched in sector opening decreased 3 times when the clearance between contactors was $\Delta\ell_{AECM} = 0,5cm$.

Chapter 4

REGULATION OF THE MODE OF DEFORMATION OF CABLE AND GUY STAYED TRUSSES USING ELECTROMECHANICAL AND FIBER OPTIC SENSORS

4.1. INTRODUCTION

Cable, guy and combined systems on their basis are varied. Among them are constructions with complex surface, shape and structural design. But they are characterized with one common feature expressed in regulation of cables and guys strain, as a result of which force factor, displacements, periods and amplitudes change under the effect of different loads. Figure 4.1 shows the diagram of combined guy construction on the model of which guy strain regulation was tested in automatic regime (Author's right N 682624), while in Figure 4.2 the corresponding photos are presented. Similarly, pyramidal construction was experimentally tested with regulation of guy tension (Patent P 2540).

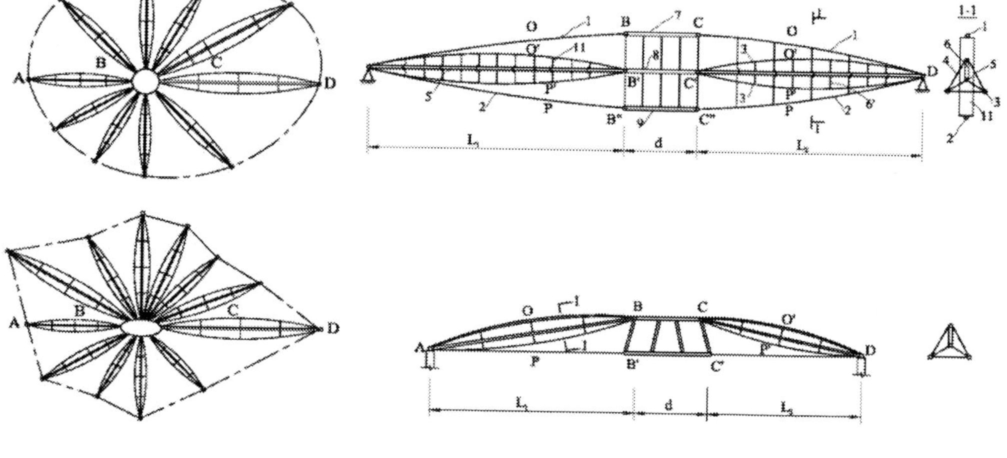

Figure 4.1. Diagrams of combined guy constructions.

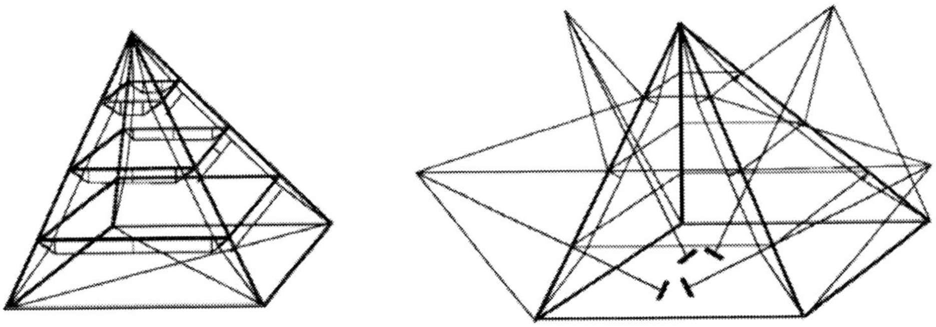

Figure 4.2. Diagrams of combined pyramidal constructions.

Figure 4.3. General view of guy tension regulation.

4.2. Regulation of Seismic Load in One- and Two-band Cable Systems

The Standards acting in Georgia indicate that vertical seismic load is to be considered in spatial floors with spans of 24 m and more. In a number of monographs the calculation of suspended large span floors is considered for vertical and horizontal seismic actions.

Seismic vibrations for large span constructions when seismic wave length is commensurable to structure dimensions in plan, as well as, seismic load determination in suspended floors are considered.

Consider multi-span filament-string for variable rigidity stretch and with identical spans equal to l. Assume that running seismic wave is of sinusoidal shape and moves with constant speed. Seismic wave motion causes horizontal and vertical vibrations of roof supports that are transmitted to the suspended system (Table 4.1, Figure 4.3).

Table 4.1.

N N n/n	Diagram of effect	Mode number	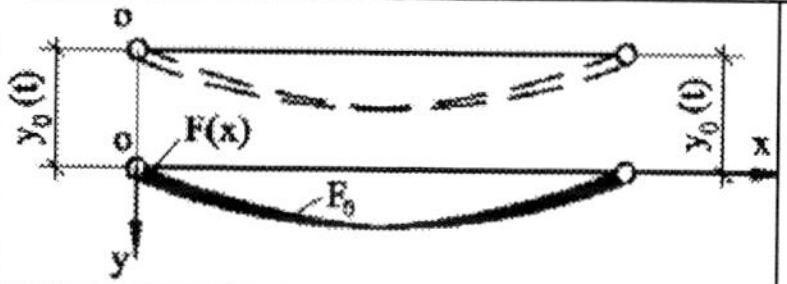 	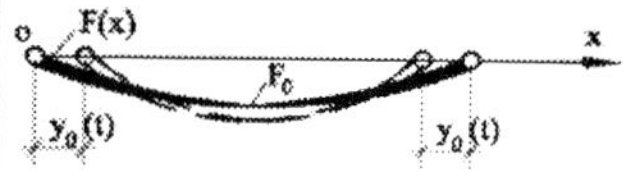
1	Fool load $r(x_1 t)$ Coefficient of mode $\Im_1(x)$ Load $r_i(x,t)$ at correspondin g vibration periods	I mode Vibration mode $\sin\frac{\pi}{\ell}x$	$-m(x)\ddot{y}_0(t) = -\rho F(x)\ddot{y}(t)$ $\eta_1 = \frac{6(2\ell^2+3f^2)}{\pi(3\ell^2+8f^2)}\sin\frac{\pi}{\ell}x$ $r_1(x,t) = \frac{6(2\ell^2+3f^2)}{\pi(3\ell^2+8f^2)}\rho F_0$ $\left[1+\frac{16f^2}{\ell^2}\left(1-\frac{2}{\ell}x\right)^2\right]^{1/2}$ $\ddot{y}(t)\sin\frac{\pi}{\ell}x$ $T_1 = 1/\sqrt{\frac{3H}{\rho F_0(3\ell^2+8f^2)}}cek$	$-m(x)\ddot{y}(t)\frac{4}{\pi^2}\frac{\ell}{f}\sin\frac{\pi}{\ell}x =$ $= -\rho F(x)\ddot{y}_0(t)\frac{4}{\pi^2}\sin\frac{\pi}{\ell}x$ $\eta_1=1$ $r_1(x,t) = \frac{4}{\pi^2}\frac{\ell}{f}\rho F_0$ $\left[1+\frac{16f^2}{\ell^2}\left(1-2\frac{x}{\ell}\right)^2\right]^{1/2}$ $\ddot{y}_0(t)\sin\frac{\pi}{\ell}$
2		I mode, square parabola	$\eta_1 = \frac{7(5\ell^2+8f^2)}{(7\ell^2+8f^2)}\frac{x}{\ell}\left(1-\frac{x}{\ell}\right)$ $T_1 = 2\pi/\sqrt{\frac{70H}{\rho F_0(7\ell^2+8f^2)}}cek$	$r_1(x,t) = \frac{16}{3}\frac{f^2}{\ell}\rho F_0$ $\left[1+\frac{16f^2}{\ell^2}\left(1-2\frac{x}{\ell}\right)^2\right]^{1/2}$ $\ddot{y}_0(t)\sin\frac{4x}{\ell}\left(1-\frac{x}{\ell}\right)$
3		II mode	$\eta_2=0$ $r_2(x,t)=0$	$\eta_2=0$ $r_2(x,t)=0$
4	.	III mode Vibration mode $\sin\frac{3\pi}{\ell}x$	$\eta_3 = \frac{2(18\ell^2+131f^2)}{9\pi(3\ell^2+8f^2)}\sin\frac{3\pi}{\ell}x$ $r_3(x,t) = \frac{2(18\ell^2+131f^2)}{9\pi(3\ell^2+8f^2)}\rho F_0$ $\left[1+\frac{16f^2}{\ell^2}\left(1-2\frac{x}{\ell}\right)^2\right]^{1/2}$ $\ddot{y}(t)\sin\frac{3\pi}{\ell}x$ $T_3 = 2/3\sqrt{\frac{3H}{\rho F_0(3\ell^2+8f^2)}}$	$r_3(x,t)=0$ $\eta_3=0$

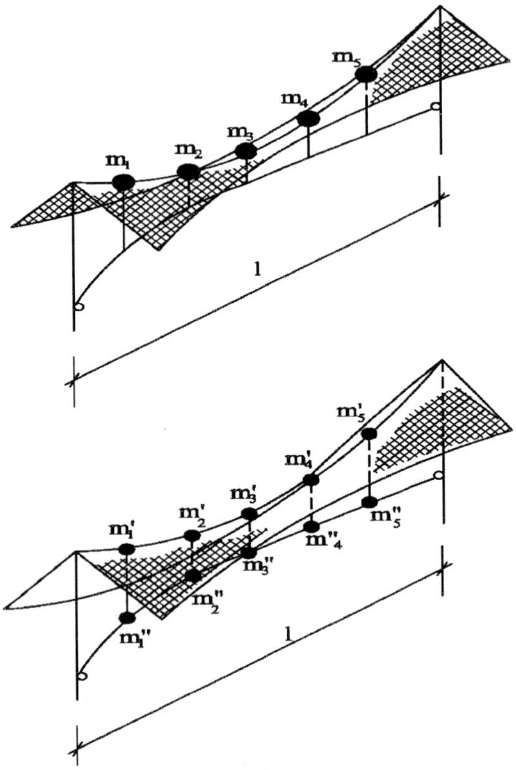

Figure 4.4. Horizontal and vertical vibrations.

Vibrations of the suspended system are caused by inertial forces. Inertial load for multi-span systems is the same as that for single-span one (Figure 4.4).

The expression of seismic load has the following form:

$$S_i(x) = q(x) \cdot k_c \cdot \mu(x) \cdot \eta_i(x) \cdot \beta_i \tag{4.1}$$

Here q(x) is static distributed load:

$$k = \frac{\ddot{y}_0(t)}{g}$$

where g is free fall acceleration;

$\ddot{y}(t)$ is maximum acceleration of soil under supports;

$\mu(x)$ is the function of transportation motion distribution within the span;

$\eta_i(x)$ is the coefficient of vibration types defined by:

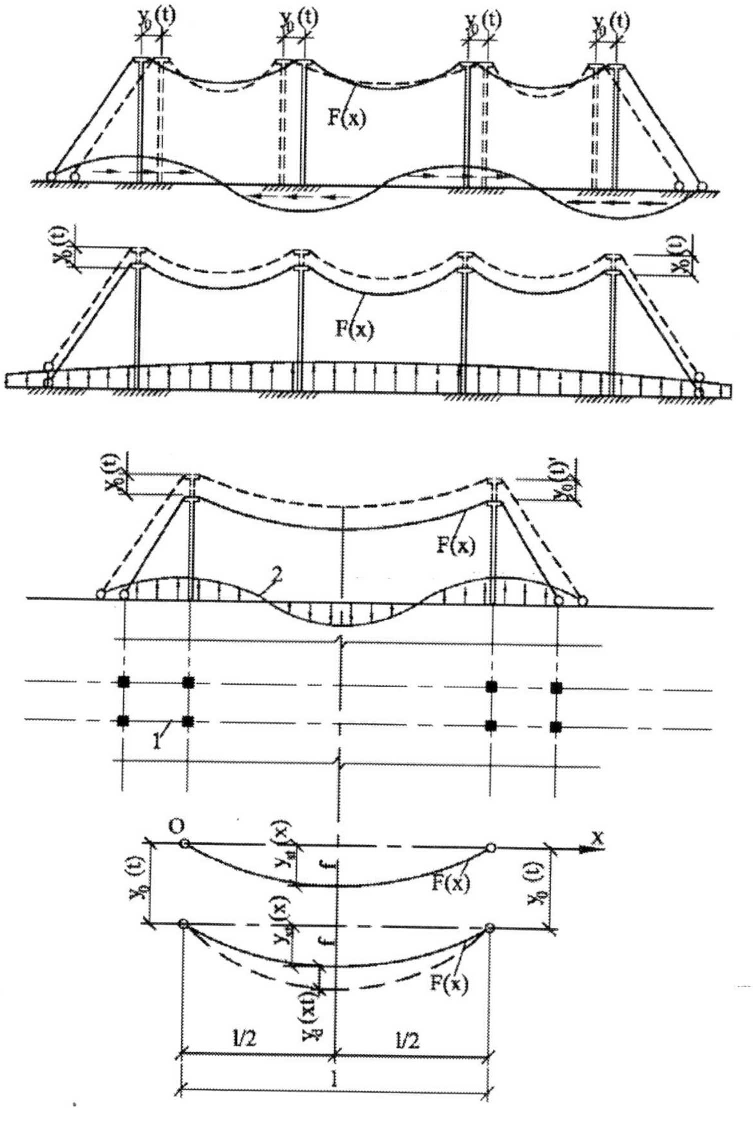

Figure 4.5. Diagram of multi-span systems under inertial forces.

$$\eta_i(x) = \frac{r_i(x,t)}{r(x,t)} = \frac{x_i(x)\int_0^\ell m(x)\cdot\mu(x)\cdot X_i(x)dx}{\mu_i(x)\int_0^\ell m(x)\cdot X_i^2(x)dx} \qquad (4.2)$$

where $X_i(x)$ is some known function characteristic for system's corresponding vibration modes and satisfying boundary conditions;

$r_i(x,t)$ is external inertial load in i direction;

$r(x,t)$ is entire external inertial load;

$\beta_i = \frac{\alpha}{T_i}$ is the coefficient of dynamism;

T_i is the period of i mode vibrations;

m(x) is variable mass value equal to ρF(x);

ρ is material density, ρ=γ/g;

γ is material specific weight;

g is free fall acceleration;

F(x) is cable-string cross-section area.

The diagrams of loading with inertial forces depend on the relation of acting seismic wave length to structure spans.

If we consider cable-string of variable rigidity or density which changes according to the following law:

$$F(x) = F_0 \left[1 + \frac{16 f^2}{\ell^2} \left(1 - 2\frac{x}{\ell} \right)^2 \right]^{1/2} \tag{4.3}$$

then the greatest strain in filament-string will occur when there happen one and a half wave vibrations, as the vibrations of just this mode are attended with cable-string lengthening while other modes of vibrations cause mainly the change of sag of cable-string.

For the case when supports move horizontally to each other the increase or decrease of sag for Δf cause the change which depends on convergence and divergence of supports by value $2y_0$, i.e. sag change to span change is:

$$2y_0 = \Delta\ell = \frac{1}{2}\int_0^\ell (y')^2\,dx \tag{4.4}$$

If sag mode is sinusoidal then:

$$2y_0 = \Delta\ell = \frac{\pi^2 f^2}{4\ell}\Delta f \tag{4.5}$$

or if sag mode is square parabola then:

$$2y_0 = \Delta\ell = \frac{16 f^2}{3\ell}\Delta f \tag{4.6}$$

Consider one- and two-band cable systems: elastic string and cable girder which have n localized mass in joints.

Span l = 6.0 m;

Sag f=0.1 m;

$E_\gamma F_\gamma$=4122, 3kN (412230 kg); H_q=0,175kN (17.5 kg);

m_k= 1 kg m/sec^2=0,01kN cm/sec^2.
For two-band cable truss the value of prestressing is:

H_l=1,6kN (160 kg).

Characteristic equation of flexible string will be:

$$\begin{vmatrix} 2,063-\lambda & 0,493 & -0,498 & -0,911 & -0,745 \\ 0,432 & 2,662-\lambda & 0,049 & -0,765 & -0,782 \\ -0,394 & 0,068 & 1,386-\lambda & 0,068 & -0,394 \\ -0,782 & -0,765 & 0,049 & 1,662-\lambda & 0,432 \\ -0,745 & -0,911 & -0,498 & 0,493 & 2,063-\lambda \end{vmatrix} = 0$$

For two-band cable truss we shall have:

$$\begin{vmatrix} 0,16641-\lambda & 0,06218 & -0,006772 & -0,04033 & -0,03862 \\ 0,06096 & 0,18029-\lambda & 0-0,5083 & -0,02279 & -0,04058 \\ -0,00746 & 0,05051 & 0,17105-\lambda & 0,05051 & -0,00746 \\ -0,04058 & -0,02279 & 0,05083 & 0,18029-\lambda & 0,06096 \\ -0,03862 & -0,04033 & -0,00672 & 0,06218 & 0,16641-\lambda \end{vmatrix} = 0$$

Where $\lambda_i = \frac{1}{\omega_i^2}$; here ω_i^2 is the square of natural vibrations frequency.

As a result of solution of characteristic equations on PC we get periods and modes of vibrations that are given in Figure 4.6. As the analysis of the received results show for flexible string the bunton caused by seismic forces is H_s = 0.21kN (21.18 kg), by sleet and wind – H_W+H_q=1.76 kN (176.25 kg); deformations are γ_{maxS}=0,057 cm and γ_{max}=28,61 cm, respectively.

From the results received for single-band and two-band cable systems we can deduce the following: a) for variable rigidity cable-string the gain of dynamic bunton is possible at vertical seismic vibrations when supports move in one phase and vibrations have the mode of one half-wave sinusoid. b) At horizontal seismic action the supports move to each other inside and outside, simultaneously.

Seismic power can be regulated by changing of natural vibrations period of thread- string and depends on f/ℓ relation.

The gain of dynamic bunton for cable girder with discrete masses is given by the second type of vibrations and for elastic string - the fifth form.

Because of closeness of natural vibration periods of the third and fourth types of cable girder vibrations have unstable form.

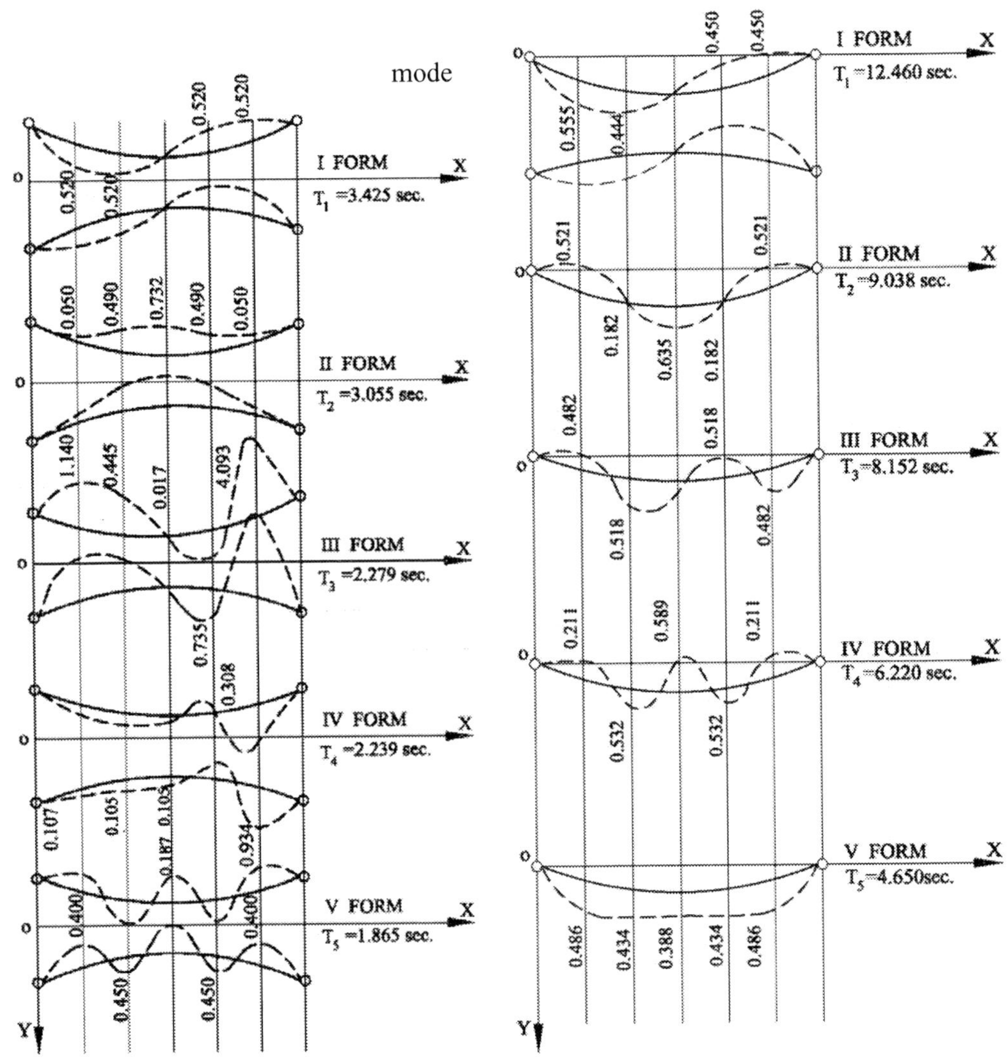

Figure 4.6. Computer calculated vibration modes.

4.3. Regulation of Membrane Aero-Elastic Stability in Wind Flow

The vibrations that can lead to the loss of dynamic stability in certain conditions appear in the steel membrane of the buildings floors under nonstationary wind flow action. These membranes vibration sources are the pulsation component of wind flow, as well as Karman vortexes stalls from the badly flow-around edges of the index contour.

Let wind velocity vector V be parallel to plane XOY and coincide with positive direction of axis X (Figure 4.1.). The equation of membrane vibrations can be written as:

$$L(\overline{\omega},\Phi)+\overline{m}\frac{\partial^2\overline{\omega}}{\partial t^2}+2\varepsilon\overline{m}+R\overline{\omega}=P(x,y)S(t) \tag{4.7}$$

where operator $L(\overline{\omega},\Phi)=-T_1\frac{\partial^2\overline{\omega}}{\partial x^2}-2T_{12}\frac{\partial^2\overline{\omega}}{\partial x\partial y}-T_2\frac{\partial^2\overline{\omega}}{\partial y^2}$

Here T_1, T_{12}, T_2 are normal and tangent forces concerning the membrane section length unit;

K is elastic foundation yielding;

P(x,y)S(t) is disturbance force.

The last expression can be represented as the sum of Karman vortexes P_k and wind pulsation P_p.

$$P_k(x,y)S(t)=\frac{1}{2}\rho V^2\cdot F\cdot C_R\cdot\sin\theta t=P_s\cdot\sin\theta t$$

$$P_p(x,y)S(t)=\sum_{i=1}^{i=n}\int_0^{\tau}\frac{1}{2}\rho V^2\cdot F\cdot C_R\cdot dt=\sum_{i=1}^{i=n}\int_0^{\tau}P_s dt$$

where ρ is air density;

F is membrane windward area;

C_R is full aerodynamic coefficient;

t - is time;

θ - is Karman vortexes stall frequency.

It's possible to determine the membrane roof total aerodynamic coefficient if the distribution of pressure coefficient along its profile is known:

$$C_R=\sqrt{C_x^2+C_y^2}$$

If we neglect the tangential force then $C_R\approx C_x$ and the last values obtained in the result of wind tunnel tests of different type membrane models are statistically processes and tabulated.

Pulse shape of the gusts can be approximated as the triangles (Figure 4.2) and then the right part of the equation will be:

$$P(x,y)S(t)=P_s\cdot X$$

where x is tabular coefficient and $P_6=\frac{1}{2}\rho V^2\cdot C_R\cdot F$.

Contour forces T_x and T_y can be defined as for the elastic band of the unit width in accordance with I.Ya.Shtaerman formula:

$$T_x = \frac{q\ell_x^2}{2\left(\sqrt{h_1} \pm \sqrt{h_2}\right)^2}$$

$$T_y = \frac{q\ell_y^2}{8f}$$

Membrane roof in the shape of buckled polygon in plan is considered in general view, particularly, it may be the rectangular membrane with arbitrary forces of thrust along the contour. Let us consider, as the design model, the version when contour forces in membrane in the direction of axis X are constant a long the length T(y) = const and they are variable in the direction of axis y, according to the law:

$$T(x) = T_{0x}\left[1 + \left(\frac{\alpha y}{\ell_y}\right)^{\mu}\right]$$

is numerical coefficient, μ=0,1,2,3,...

Taking the membrane vibration shape as $\overline{\omega}(x,y,t) = Z(t)W(x)V(y)$ and using Bubnov-Galerkin procedure we shall obtain:

$$\ddot{z} + 2\varepsilon z + \omega^2 z = P(x,y)S(t)\frac{R_7}{\overline{m}R_5} \tag{4.8}$$

where z(t) is displacement;

ε is damping coefficient;

ω^2 is natural oscillations frequency square which with μ=1 is determined according to formula:

$$\omega = \pi\sqrt{T_{ox}\frac{n^2}{\overline{m}\ell_x^2}\left(1 + \frac{\alpha}{2}\right) + T_{oy}\frac{m^2}{\overline{m}\ell_y^2} + \frac{k}{\overline{m}}} \tag{4.9}$$

and with μ=2

$$\omega = \pi\sqrt{T_{ox}\frac{n^2}{\overline{m}\ell_x^2}\left(1 + \frac{\alpha^2}{3}\right) + T_{oy}\frac{m^2}{\overline{m}\ell_y^2} + \frac{k}{\overline{m}}} \tag{4.10}$$

$\overline{m}$ is membrane surface unit mass

$$R_7 = \int_0^{\ell_x}\int_0^{\ell_y} W(x)V(y)dxdy; \; R_5 = \int_0^{\ell_x}\int_0^{\ell_y} W^2(x)V^2(y)dxdy$$

In particular case when $T_{0x} = T_{0y} = \overline{S}$, k=0 we obtain known formulas of the square membrane natural oscillations frequency.

When the number of pulses is large, for instance, if it exceeds 0.5/γ where γ is the coefficient of the membrane material internal friction and changes within ranges γ=0,01÷0,025, then the largest in time displacement of the roof is found as:

$$Z_0^* = Z_0 \cdot \psi \tag{4.11}$$

where Z_0 is displacement induced by the momentary pulse

$$Z_0 = \frac{\varepsilon_1 \cdot S}{\overline{m}_1 \cdot \omega_1} e^{-\frac{\gamma\pi}{4}}$$

ε_1 is determined in diagram or table depending on pulse shape and

$$\tau^* = \frac{\tau}{T_i}.$$

$S = P_s \cdot \tau$ τ -is duration of gust;

ω_1 is natural oscillations circular frequency in rad/sec.

$$\psi = \frac{\sqrt{1 + e^{2\pi\gamma\overline{\theta}} - 2 \cdot e^{\pi\gamma\overline{\theta}} \cos(2\pi\overline{\theta})}}{2ch(\pi\gamma\overline{\theta}) - 2\cos(2\pi\overline{\theta})}$$

$\overline{\theta} = \dfrac{T_0}{T_n}$; where T_0 is gust period.

Proceeding from the expression for Strouhal quantity and coincidence condition of the vortexes stall frequency with the natural oscillations frequency of the membrane roof the following equality can be written:

$$Sh = \frac{2\pi \cdot \ell}{V \cdot T}$$

Hence the expression for wind critical velocity for membrane roof will be written as:

$$V_{CR} = \frac{2\pi\ell}{SR \cdot T}$$

Knowing the Strouhal quantity value Sh=0,1÷0,3, it can be supposed that critical velocities correspond to the actual wind velocities only at very large membrane spans of about some hundred meters.

Really, for the trapezoidal in plan membrane roof of the musical theatre in Tbilisi with the span of 30 m, thickness of 4 mm and supported by the orthogonal bands system, critical velocity of the aero-elastic stability loss turns out to be less than 150 m/sec.

Natural oscillations frequencies of the same membrane calculated according to the above cited formulas make up W_{11} = 0.94 Hz and W_{22}=1.81 Hz, that conforms well to the values calculated according to the formulas suggested by other authors.

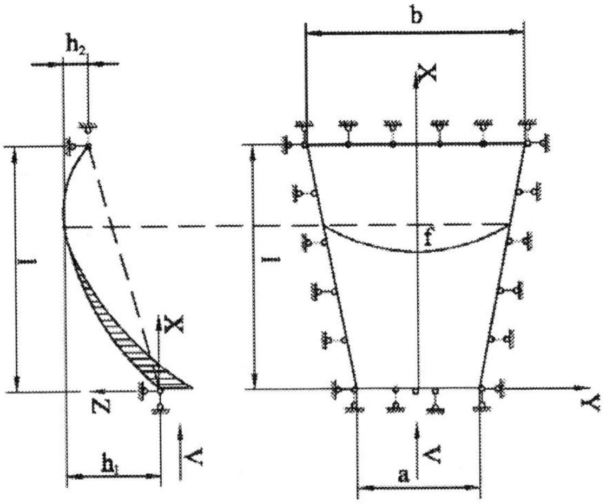

Figure 4.7. Geometrical shape of membrane.

4.4. Regulation of Vibrations of Circular Contour Plane Membrane on Suspended Cable

The equation of circular contour plane membrane motion is:

$$\frac{\partial^2\omega}{\partial r^2}+\frac{1}{r}\frac{\partial\omega}{\partial r}+\frac{1}{r^2}\frac{\partial^2\omega}{\partial\varphi^2}-\frac{\rho}{H}\frac{\partial^2\omega}{\partial t^2}-\frac{P_r(t)}{H}\frac{\partial^2\omega}{\partial r^2}-\frac{P_\varphi(t)}{H}\left(\frac{1}{r^2}\frac{\partial^2\omega}{\partial\varphi^2}+\frac{1}{r}\frac{\partial\omega}{\partial r}\right)+\frac{k(r,\varphi)}{H}\omega=B(t) \tag{4.12}$$

If we consider only natural vibrations and take function:

$$\omega(r,\varphi,t)=f(t)\psi(r,\varphi) \tag{4.13}$$

using Bubnov-Galerkin method we shall receive Mathieu type equation:

$$\frac{d^2 f(t)}{dt^2}+\omega_n^2\left(1-\mu\cos\theta t\right)f(t)=0 \tag{4.14}$$

Here we have: $P_r(t) = P\varphi(t) = q_0 + q_t \cos\theta t$

ω_n^2 is natural vibrations frequency:

$$\omega_n^2 = -(R_1 + R_2 + R_3 + R_4) / R_5 \tag{4.15}$$

μ is pulsation coefficient:

$$\mu = \frac{q_t}{H}(R_1 + R_2 + R_3) / (R_1 + R_2 + R_3 + R_4) \tag{4.16}$$

Here quadratures are:

$$\begin{aligned}
R_1 &= \int_0^R \int_0^{2\pi} \frac{d^2\psi}{dr^2} \psi \, dr d\varphi \\
R_2 &= \int_0^R \int_0^{2\pi} \frac{1}{r} \frac{d\psi}{dr} \psi \, dr d\varphi \\
R_3 &= \int_0^R \int_0^{2\pi} \frac{1}{r^2} \frac{d^2\psi}{d\varphi^2} \psi \, dr d\varphi \\
R_4 &= \int_0^R \int_0^{2\pi} \frac{k(r,\varphi)}{H} \psi^2 \, dr d\varphi \\
R_5 &= \int_0^R \int_0^{2\pi} \frac{\rho}{H} \psi^2 \, dr d\varphi
\end{aligned} \tag{4.17}$$

Here we can receive the following approximating functions that satisfy boundary condition and are depended only on radius:

$$\begin{aligned}
\psi(r) &= c(r^2 - R^2) = c_0 \left[\left(\frac{r}{R} \right)^2 - 1 \right]; \quad c_0 = cR^2 \\
\psi(r) &= c_1 \left[\left(\frac{r}{R} \right)^2 - 1 \right]^2 + c_2 \left[\left(\frac{r}{R} \right)^2 - 1 \right]^3 + \cdots \\
\psi(r) &= a_1 \cos\frac{\pi r}{2R} + a_2 \cos\frac{3\pi r}{2R} + \cdots
\end{aligned} \tag{4.18}$$

The flexibility of membrane elastic base can be:

$$k(r,\varphi)=k_0=const$$
$$k(r,\varphi)=k_0\delta(r-r_0)\delta(\varphi-\varphi_0) \qquad (4.19)$$
$$k(r,\varphi)=k_0\frac{4r}{R^2}(R-r)\delta(\varphi-\varphi_0)$$

Here δ is Dirack function.

If we admit that circular contour plane surface membrane vibrates only in radial direction i.e., vibrations happen when ω depends only on r and t such vibrations may occur if the initial conditions are as follows:

$$\omega\big|_{t=0}=\beta_0(r) \qquad \frac{\partial\omega}{\partial t}\bigg|_{t=0}=\beta_1(r)$$

where $\beta_0(r)$ and $\beta_1(r)$ are given functions in interval (O, R).

As in the mentioned case ω does not depend on angle φ the membrane motion equation is simplified and is as follows:

$$\frac{\partial^2\omega}{\partial r^2}+\frac{1}{r}\frac{\partial\omega}{\partial r}-\frac{\rho}{H}\frac{\partial^2\omega}{\partial t^2}-\frac{P_r(t)}{H}\frac{\partial^2\omega}{\partial r^2}-\frac{P_\varphi(t)}{H}\frac{1}{r}\frac{\partial\omega}{\partial r}+\frac{k(r)}{H}\omega=0 \qquad (4.20)$$

If we admit that $P_r(t)=q_t\cos\theta t$ and use Bubnov-Galerkin method we shall receive Mathieu type equation:

$$\frac{d^2\overline{f}}{dt^2}+\overline{\omega}_n^2\left(1-\overline{\mu}\cos\theta t\right)\overline{f}=0 \qquad (4.21)$$

where $\overline{\omega}_n^2$ is the frequency of membrane natural vibration:

$$\overline{\omega}_n^2=-\left(\overline{R}_1+\overline{R}_2+\overline{R}_4\right)/\overline{R}_5 \qquad (4.22)$$

Here:

$$\overline{R}_1=\int_0^R\frac{d^2\psi}{dr^2}\psi dr, \quad \overline{R}_2=\int_0^R\frac{1}{r}\frac{d\psi}{dr}\psi dr,$$
$$\overline{R}_4=\int_0^R\frac{k(r)}{H}\psi^2dr, \quad \overline{R}_5=\frac{\rho}{H}\int_0^R\psi^2dr, \qquad (4.23)$$

$\overline{\mu}$ is pulsation coefficient:

$$\bar{\mu} = \frac{q_t}{H}\left(\bar{R}_1 + \bar{R}_2\right) / \left(\bar{R}_1 + \bar{R}_2 + \bar{R}_4\right) \tag{4.24}$$

Let approximating function be:

$$\psi(r) = c(r^2 - R^2) + \cdots \tag{4.25}$$

and elastic base coefficient:

$$k(r) = k_0 \delta(r - r_0) \tag{4.26}$$

If we insert (4.22) and calculate integrals, we shall receive:

$$\bar{\omega}_1 = \sqrt{\frac{5H}{\rho R^2} + \frac{15}{8}\frac{k_0}{\rho}\frac{r_0^4}{R^5} - \frac{15}{8}\frac{k_0}{\rho}\frac{r_0^2}{R^3} + \frac{15}{8}\frac{k_0}{\rho}\frac{1}{R}} \tag{4.27}$$

Let consider a particular case when $k_0=0$, then:

$$\bar{\omega}_1 = \sqrt{\frac{5H}{\rho R^2}} = \frac{2,236}{R}\sqrt{\frac{H}{\rho}} \tag{4.28}$$

The exact value is:

$$\bar{\omega}_1 = \frac{2,4048}{R}\sqrt{\frac{H}{\rho}} \tag{4.29}$$

When elastic support is in the centre of membrane or $r_0=0$, then

$$\bar{\omega}_1 = \sqrt{\frac{5H}{\rho R^2} + \frac{15}{8}\frac{k_0}{\rho}\frac{1}{R}} \tag{4.30}$$

When elastic support is on the contour, then $r_0=R$ and $\bar{\omega}_1$ is determined with formula (4.28).

When flexible support is in the quarter of diameter - $r_0 = \frac{R}{2}$, then:

$$\bar{\omega}_1 = \sqrt{\frac{5H}{\rho R^2} + \frac{135}{128}\frac{k_0}{\rho R}} \tag{4.31}$$

Chapter 5

REGULATION OF STRESSES AND STRAINS IN SPATIAL COMPOSITE CONSTRUCTIONS WITH ELECTRO-MECHANICAL AND FIBER-OPTICAL SENSORS

5.1. INTRODUCTION

In different spheres of building and technique spatial composite constructions including composite slabs and shells are widely used. They differ from usual spatial constructions as cable or guy systems are used there.

Conditions of shells supported on four sides greatly determine their stressed-strained state.

As a rule arcs and trusses are used as profile constructions for shells and trusses and the effect of their yielding on working conditions of shells is considered.

If shell profile or any point of its surface is supported on cable or guy system their co-working is discussed which effects stressed-strained state of shell body.

Prestressing of cable or guy systems or force regulation in them causes the yield changes in shell profile and body in respective local zones, similarly concentrated external forces applied on a small section cause shell bending in limited area and its value depends on shell curvature, its thickness, geometry in plan and other parameters.

The mentioned effect of concentrated force is considered for a cylindrical shell, in the form of composite system, as tie guy and cable are used. Here are considered some problems of regulation of composite spatial constructions.

5.2. REGULATION OF FORCES IN CYLINDRICAL SHELLS

For regulation of forces, particularly of bending moment, open profile shell with local concentrated loads is used in cylindrical shells. Shell materials was steel Cm3, with Poisson coefficient μ=0,3, modulus of flexibility E =$2{,}1 \cdot 10^6$kg/cm^2 or $2{,}1 \cdot 10^4$kN/cm^2. Shell dimensions in plan are B = 2000 mm (length), span l =1100 mm; shell thickness δ= 2.5 mm; cross section area boom f =290 mm; radius R = 66.2 cm.

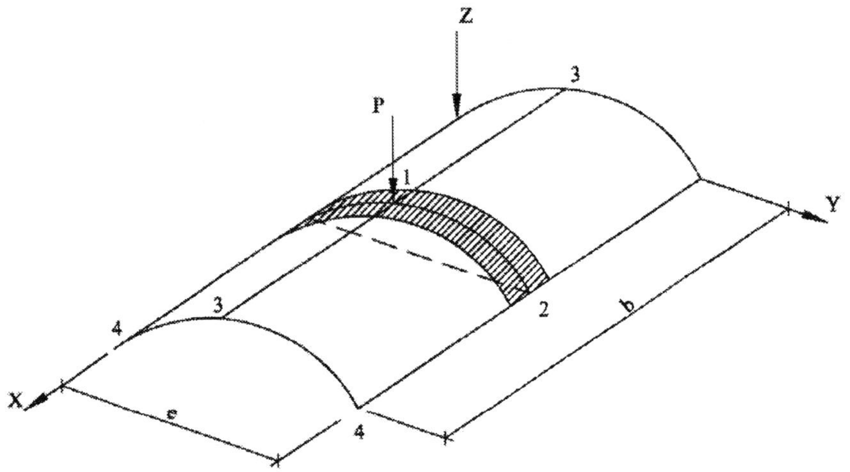

Figure 5.1. Design diagram of composite cylindrical shell.

In the middle of cylindrical shell length tie - guy is arranged which is prestressed with force H = 0.06 kN.

For regulation of bending moment in cylindrical shell on tie top in shell body a unit width strip is taken as given in Figure 5.1.

From reference book (Directory for designer: calculation and theoretical part. Ed A.A. Ushanski, M., 1960, p. 435, Table 8.2.7) it is known that:

Force in tie as a result of self stress is determined with formula:

$$X_{ss} = H_2 = \frac{25}{128} PK \frac{\ell}{f} \tag{5.1}$$

Bending moment M_l under concentrated force is:

$$M_1 = \left(0,25 - \frac{25}{128} K\right) P\ell$$

where $k = {}^{1}\!/_{(1+\nu)}$; $\nu = \frac{15}{8} \frac{D}{f^2} \left(\frac{E}{E_c F_c} + \frac{n}{\ell} \right)$.

Here $D = {}^{E\delta^3}\!/_{12(1-\mu^2)}$; n=0,785 when $\frac{f}{\ell} = 0,263$.

Total strain in tie-bar is the sum of prestrained force and tie-bar self-tension force.

Load on cylindrical shell was applied stage by stage:

P=0,2; 0,4; 0,6; 0,8; 1,0 kN.

The effectiveness of force regulation in cylindrical shell is shown on the example of bending moment in the strip cut out of shell. If we take bending moment equation in different points of the strip we shall have:

$$M_{max} = |M_{min}| \tag{5.2}$$

From this condition we can define the value of tie-bar prestressing when this condition is satisfied.

In order to satisfy the condition of bending moments equality the unit width arc with tie-bar cut out from the shell is considered as statically indefinable system after the solution of which total force in brace is received that represents algebraic sum of two forces – self tension and prestressed forces.

Here the notion of self strain is taken the force in statically indefinable system under the action of the given external loads.

In order to maintain the equality of permanently bending moments in shell cross-sections on tie-bar top it is necessary that, in case of external force change, the presstressing force value in guy-rope be respectively regulated which will be provided with automatic electro-mechanical (AECM) or fiber-optical (ABOM) system.

If we use technical literature (A.R.Rzhanitsin, Structural mechanics, M., Higher school, 1982, 130 p) self-tension force will be:

$$X_{ss} = 0,785P \doteq 0,471kN = 47,1kg$$

Bending moment caused by unit force in guy-rope in the arc strip cut out from the shell is:

$$M_1 = -f\left(1 - \frac{4x^2}{\ell^2}\right) \tag{5.3}$$

The variation of bending moment in the arc strip is:

$$M(x) = \frac{P\ell}{4}\left(1 - \frac{2x}{\ell}\right) \tag{5.4}$$

When $x = \frac{\ell}{2}$ in support points (in tie-bar attachment unit) then:

$$M\left(\frac{\ell}{2}\right) = \frac{P\ell}{4}\left(1 - \frac{2}{\ell}\cdot\frac{\ell}{2}\right) = 0 \tag{5.5}$$

X=0; the point under outer force application is:

$$M_0 = \frac{P\ell}{4}\left(1-\frac{2\cdot 0}{\ell}\right) = \frac{P\ell}{4};\ \mathrm{H}_2 = 0,785P \tag{5.6}$$

Total bending moment in arc strip is $M(x) = H_2 f(x) = H_2 f\left(1-\frac{4x^2}{\ell^2}\right)$

$$M(x) = M_p + X_{CH}\cdot M_1 = \frac{P\ell}{4}\left(1-\frac{2x}{\ell}\right) - 0,785P\cdot f\left(1-\frac{4x^2}{\ell^2}\right) \tag{5.7}$$

Maximum positive moment when x=0 is:

$$M_{\max} = \frac{P\ell}{4} - H_2 f \tag{5.8}$$

If we assume the condition $\frac{dM(x)}{dx} = 0$, we shall have;

$$\frac{dM(x)}{dx} = -\frac{P}{2} + \frac{6,28Pf}{\ell^2}x = 0$$

Hereof we receive $x = \frac{\ell^2}{12,56f^2} = 0,3\ell$ when $\frac{\ell}{f} = 3,79$.

If we introduce the received results in condition $M_{\max} = |M_{\min}|$ we shall get total force in guy-rope when x=0,3ℓ

$$\frac{P\ell}{4} - H_f\cdot f = \frac{P\ell}{4}\left(1-\frac{2\cdot 0,3\ell}{\ell}\right) - 0,785\cdot p\cdot f\left(1-\frac{4\cdot 0,3^2\cdot \ell^2}{\ell^2}\right) =$$

$$= \frac{P\ell}{4}(1-0,6) - 0,785p\cdot f(1-0,36) = \frac{P\ell}{4}\cdot 0,4 - 0,785\cdot p\cdot f\cdot 0,64$$

Hereof $\frac{P\ell}{4} - 0,4\frac{P\ell}{4} + 0,5pf = H_f\cdot f$ or $0,6\frac{P\ell}{4} + 0,5p\cdot f = H_f\cdot f$

We receive $\frac{3}{20}\frac{P\ell}{f} + \frac{1}{2}p = H_f$; $\frac{\ell}{f} = 3,79$ then $0,57P+0,5P=H_f$

in the end $H_f=1,07p=1,07\cdot 0,6=0,64kn=64,0kg$

as $X_{ss}=0,785P=0.471kn$
Then prestressing force will be self-tension force abstracted from total force or:

$H_{PS}=H_f-X_{SS}=0,64-0,471=0,169kn=16,9kg$

Then under the given external load P=60kg (0,6kN) in order to satisfy the condition $M_{\max} = |M_{\min}|$ the value of prestressing force in guy-rope will be:

$$H_{ps} = H_f - X_{ss} = 0,64 - 0,471 = 0,169kn = 16,9kg$$

or in other way:

$$H_{ps} = p\left(\frac{\ell}{6f} + 0,436\right) - 0,785p = p\left(\frac{3,79}{6} + 0,436\right) - 0,785p = P(0,631 + 0,436) - 0,785p =$$
$$= p(0,631 + 0,436 - 0,785) = 0,282p$$

When P =60kg (0,6kN) then H_{ps}=16,92kg (0,169kN).

If the total force in a tie-bar is H_f =0,64kN (64kg) and tie-bar length is ℓ=100cm; cable diameter is d_{cabl}=0,25cm; cable area $F_{cabl} = \frac{\pi d^2}{4} = \frac{3,14 \cdot 0,25^2}{4} = 0,05cm^2$;

$$E_{cabl}F_{cabl} = 2,1 \cdot 10^6 \cdot 0,05 \cong 0,1 \cdot 10^6 kg = 0,1 \cdot 10^4 kN$$

parameter $\frac{\ell_{cabl}}{E_{cabl}F_{cabl}} = \frac{100}{0,1 \cdot 10^6} = 0,001$;

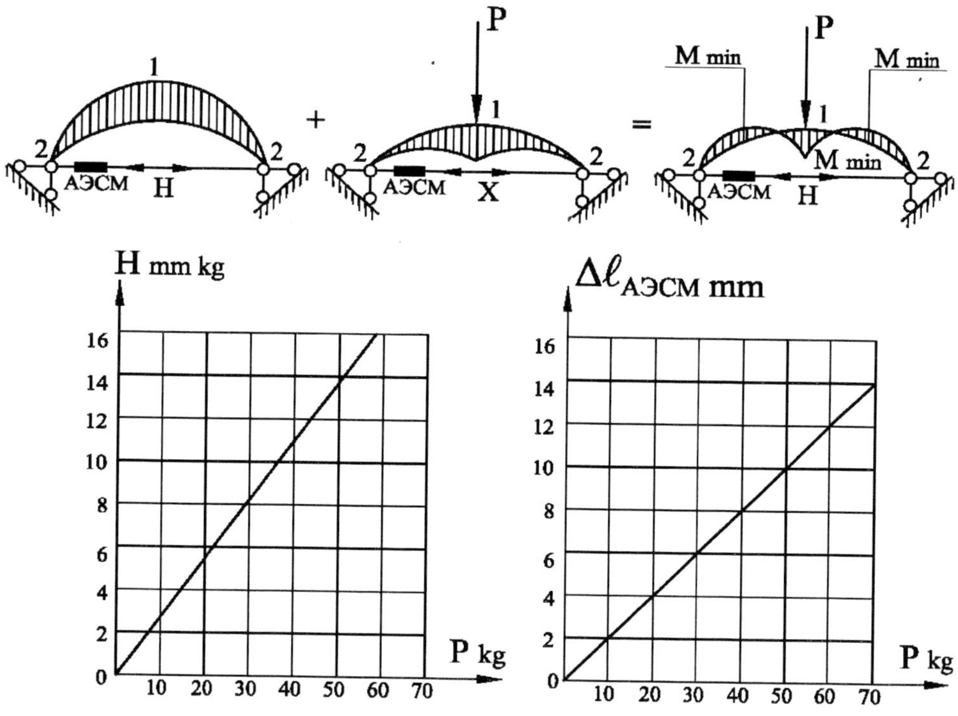

Figure 5.2. The relation of tension in tie-bar, brace-coupling gap and concentrated force.

then according to Hook law strain in tie-bar is:

$$\Delta\ell_{cabl} = H_f \frac{\ell_{cabl}}{E_{cabl}F_{cabl}} = 64 \cdot 0,001 = 0,064cm = 0,64mm$$

As a result of test we get composite system tie-bar displacement under load

Then the gap between contactors of electro-mechanical sensor is $\Delta\ell = 2 \cdot v_2 = 2 \cdot 3,572mm = 7,144sm$

$$\Delta\ell_{АЭСМ} = 7,144 - 0,64mm = \Delta\ell - \Delta\ell_{cabl} = 6,5040mm$$

Theoretically $\Delta\ell_{red}$ and y=3,324mm, then $\Delta\ell_{АЭСМ} = 2 \bullet 3,324 - 0,64 = 6,648 - 0,64 = 6,008mm$

5.3. Regulation of Strains in Cylindrical Shells

For estimation of stresses and force factors in cylindrical shells V.Z.Vlasov's technical theory for thin flexible shells is used. According to this theory using two cuts from cylindrical shell body along generatrix and two circumferences an infinitely small element is taken. Introduce dimensionless coordinates:

$$\xi = \frac{x}{R}; \quad \theta = \frac{S}{R}; \quad C^2 = \frac{h^2}{12R^2} \tag{5.9}$$

where S is arc element, C is dimensionless parameter; R is shell radius; h is shell thickness.

The displacement of middle surface points of the shell along coordinate axis x, y, and z denote as u, v and w and assume as unknown.

Then shell equation system in displacements has the following form (V.Z.Vlasov, Selection. V. I, AC USSR, 1962, p.261; P.M.Varvak, Z.P.Varvak, Net method in the problems of building constructions calculation. M.. Stroiizdat, 1977, p.122).

$$\frac{\partial^2 U}{\partial \xi^2} + \frac{1-\nu}{2}\frac{\partial^2 U}{\partial \theta^2} + \frac{1+\nu}{2}\frac{\partial^2 \upsilon}{\partial \xi \partial \theta} + \nu \frac{\partial \omega}{\partial \xi} - C^2\left(\frac{\partial^3 \omega}{\partial \xi^3} - \frac{1-\nu}{2}\frac{\partial^3 \omega}{\partial \xi \partial \theta^2}\right) = \frac{\left(1-\nu^2\right)}{Eh} R^2 X$$

$$\frac{1+\nu}{2}\frac{\partial^2 U}{\partial \xi \partial \theta} + \frac{\partial^2 \upsilon}{\partial \theta_2} + \frac{1-\nu}{2}\frac{\partial^2 \upsilon}{\partial \xi^2} + \frac{\partial \omega}{\partial \theta} - \frac{(3-\nu)}{2}\frac{\partial^3 \omega}{\partial \xi^2 \partial \theta} = \frac{\left(1-\nu^2\right)}{Eh} R^2 Y \tag{5.10}$$

$$\nu\frac{\partial U}{\partial\xi}-C^{2}\left(\frac{\partial^{3}\omega}{\partial\xi^{3}}-\frac{1-\nu}{2}\frac{\partial^{3}U}{\partial\xi\partial\theta^{2}}\right)+\frac{\partial U}{\partial\theta}-\frac{(3-\nu)C^{2}}{2}\frac{\partial^{3}\upsilon}{\partial\xi^{2}\partial\theta}+C^{2}\nabla^{2}\nabla^{2}\omega+2C^{2}\frac{\partial^{2}\omega}{\partial\theta^{2}}+C^{2}\omega+\omega=$$
$$=-\frac{(1-\nu^{2})}{Eh}R^{2}Z$$

Here x, y, and z are surface force projection on axes x, y, and z for unit surface.

If we choose cylindrical net of regular structure on cylindrical shell surface with relative dimensions of cells:

$$\frac{\Delta\xi}{\Delta\theta}=\chi,\quad \frac{1-\nu}{2}=A_{\gamma};\quad \frac{1+\nu}{2}=B_{\gamma};\quad \frac{3-\nu}{2}=C_{\gamma}$$

here y is Poisson coefficient.

The above presented equations system received for i point of net after some transformations in finite-difference expression will take the form:

$$-2\left(\frac{1}{\chi}+A_{\gamma}\chi\right)U_{i}+\frac{1}{\chi}(U_{k}+U_{\ell})+A_{\gamma}\chi(U_{m}+U_{n})+\frac{B_{\gamma}}{4}\left[(\upsilon_{q}+\upsilon_{r})-(\upsilon_{0}+\upsilon_{p})\right]+$$
$$+\left(\frac{\gamma\Delta\xi}{2\chi}+\frac{c^{2}}{\chi\Delta\xi}-\frac{c^{2}A_{\gamma}}{\Delta\theta}\right)(\omega_{k}-\omega_{\ell})-\frac{c^{2}}{2\chi\Delta\xi}(\omega_{s}-\omega_{t})+\frac{c^{2}A_{\gamma}}{2\Delta\theta}(\omega_{0}-\omega_{q}-\omega_{r}-\omega_{p})=$$
$$=\frac{1-\nu^{2}}{Eh}R\Delta\xi R\Delta\theta X_{i}$$

$$-2\left(\chi+\frac{A_{\gamma}}{\chi}\right)\upsilon_{i}+\frac{A\gamma}{\chi}(\upsilon_{k}+\upsilon_{\ell})+\chi(\upsilon_{m}+\upsilon_{n})+\frac{B_{\gamma}}{4}\left[(U_{q}+U_{r})-(U_{0}+U_{p})\right]+$$
$$+\left(\frac{\Delta\xi}{2}+\frac{C_{\gamma}c^{2}}{2\chi\Delta\theta}\right)(\omega_{n}-\omega_{m})-\frac{C_{\gamma}c^{2}}{4\chi\Delta\theta}(\omega_{r}-\omega_{0}+\omega_{p}-\omega_{q})=\frac{1-\nu^{2}}{Eh}R\Delta\xi R\Delta\theta Y_{i}$$

$$\left(\frac{6\chi}{4\theta^{2}}+\frac{6}{\chi\Delta\xi^{2}}+\frac{4\chi}{\Delta\xi^{2}}+\frac{4}{\chi\Delta\theta^{2}}-4c^{2}\chi+\frac{c^{2}\Delta\xi}{\chi}+\frac{\Delta\xi^{2}}{\chi}\right)\omega_{i}-\left(\frac{2\chi}{\Delta\xi^{2}}+\frac{2}{\chi\Delta\theta^{2}}+\frac{4}{\chi\Delta\xi^{2}}\right)(\omega_{k}+\omega_{\ell})-$$
$$-\left(\frac{4\chi}{\Delta\theta^{2}}+\frac{2\chi}{\Delta\xi^{2}}+\frac{2}{\chi\Delta\theta^{2}}-2c^{2}\chi\right)(\omega_{m}+\omega_{n})+\left(\frac{\chi}{\Delta\xi^{2}}+\frac{1}{\chi\Delta\theta^{2}}\right)(\omega_{0}+\omega_{p}+\omega_{q}+\omega_{r})+$$
$$+\frac{\chi}{\Delta\theta^{2}}(\omega_{u}+\omega_{\upsilon})+\frac{1}{\chi\Delta\xi^{2}}(\omega_{s}+\omega_{t})+\left[\frac{\gamma\Delta\xi}{2\chi}+\frac{C^{2}}{\Delta\xi\chi}-\frac{(1-\nu)c^{2}}{2\Delta\theta}\right](u_{k}-u_{\ell})-$$
$$-\frac{1}{2\chi\Delta\xi}(u_{s}-u_{t})-\frac{1-\nu}{4\Delta\theta}(u_{0}-u_{q}+u_{r}-u_{p})+\left(\frac{\Delta\xi}{2c^{2}}+\frac{3-\nu}{2\Delta\theta\chi}\right)(\upsilon_{n}-\upsilon_{m})-$$
$$-\frac{3-\nu}{4\Delta\theta\chi}(\upsilon_{r}-\upsilon_{0}+\upsilon_{p}-\upsilon_{q})=\frac{(1-\nu^{2})}{Eh}R\Delta\xi R\Delta\theta Z_{i}$$

(5.11)

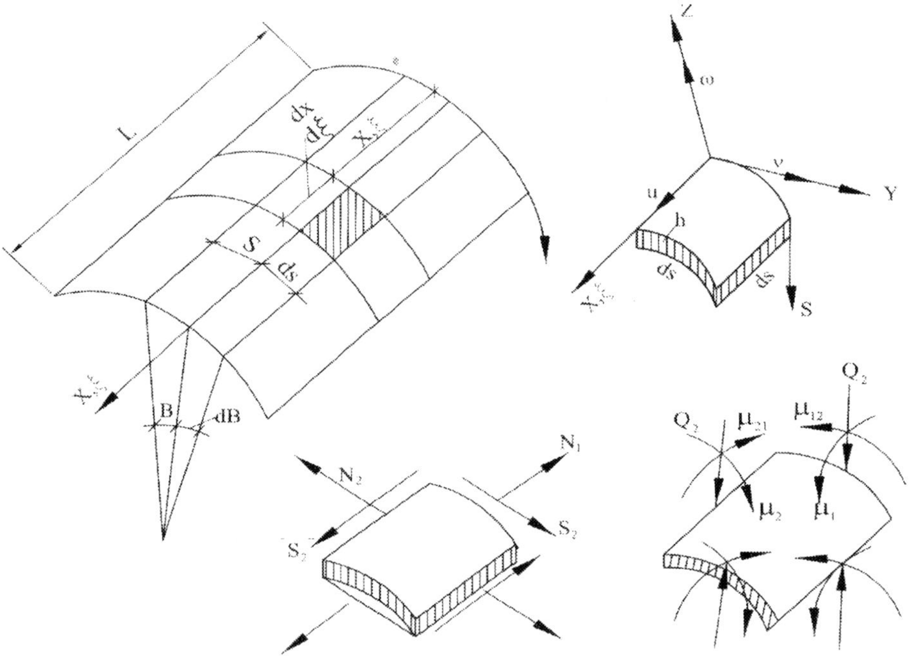

Figure 5.3. Elements of the theory of shells.

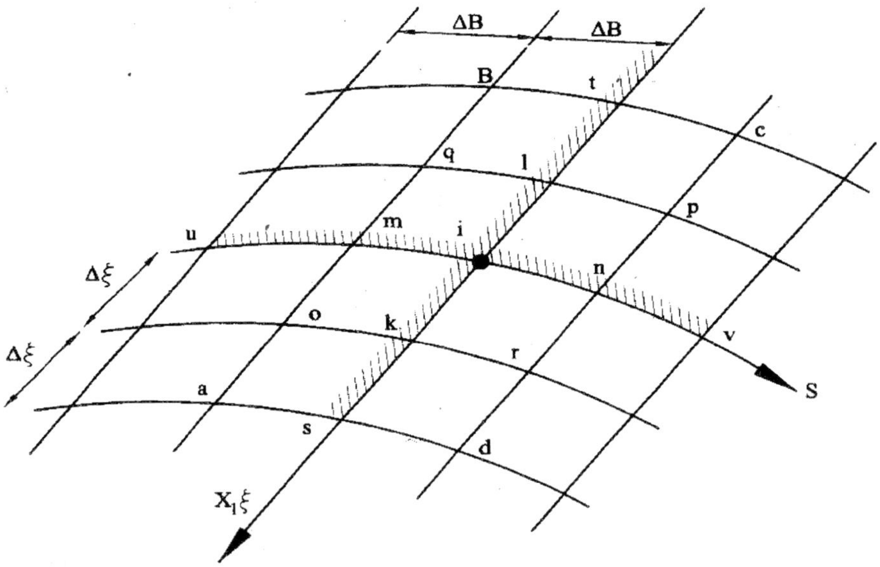

Figure 5.4. Shell surface net.

Boundary value conditions: for cylindrical shell the edges free on all four sides with fixed four angles of shell is assumed as the boundary condition.

For the points of free edges of the shell the four static conditions are valid, particularly:

$$N_2 = 0;\ M_2 = 0,\ S_2 = 0;\ Q_2^* = 0 \tag{5.12}$$

Here

$$N_2 = \frac{Eh}{(1-\nu)^2 R}\left[\gamma\frac{(u_k-u_e)}{2\Delta\xi} + \frac{(\upsilon_n-\upsilon_m)}{2\Delta\theta} + \left(1+\frac{h^2}{12R^2} - \frac{h^2}{6R^2\Delta\theta^2}\right)\omega_i + \frac{h^2}{12R^2}\frac{(\omega_m+\omega_n)}{\Delta\theta^2}\right] \quad (5.13)$$

$$S_2 = \frac{Eh}{2R(1+\nu)}\left[1+\frac{h^2}{12R^2}(u_n-u_m) + \left(1-\frac{h^2}{24R^2\Delta\xi}\right)(\upsilon_k-\upsilon_e) + \frac{(\omega_q+\omega_r)-(\omega_0+\omega_p)}{2\Delta\xi\Delta\theta}\right] \quad (5.14)$$

$$M_2 = \frac{P}{R^2}\left[\left(1-\frac{2}{\Delta\theta^2}-\frac{2\gamma}{\Delta\xi^2}\right)\omega_i + \frac{(\omega_m+\omega_n)}{\Delta\theta^2} + \gamma\frac{(\omega_k+\omega_e)}{\Delta\xi^2}\right] \quad (5.15)$$

$$Q_2^* = \frac{P}{R^3}\left\{\begin{array}{l}\frac{(1-\nu)}{2\Delta\xi^3}(\upsilon_k-2\upsilon_i+\upsilon_e) - \frac{(1-\nu)}{8\Delta\xi\Delta\theta}\left[(u_r+u_q)-(u_0+u_p)\right] + \\ + \left[\frac{1}{2\Delta\theta}+\frac{(\nu-2)}{\Delta\theta\Delta\xi^2}+\frac{(\nu-2)}{\Delta\theta^2}\right](\omega_m-\omega_n) - \frac{2(\nu-2)}{\Delta\theta^2}\omega_i\end{array}\right\} \quad (5.16)$$

For the points of angles fixation the boundary condition for point i is:

$$u_i = \upsilon_i = \omega_i = 0 \text{ and } \omega_n = \overline{\omega}_m \quad (5.17)$$

D - cylindrical stiffness is $D = \frac{Eh^3}{12(1-\nu^2)}$; E is elasticity modulus of shell material.

For experimental cylindrical shell a simple net was used in order to state vertical displacement of its three points (ω_i) in case of concentrated force application in the middle of the shell and with consideration of tie-bar placed in the middle part of shell length.

The initial parameters have been preliminarily determined, particularly:

$\chi = \frac{\Delta\xi}{\Delta\theta} = \frac{100}{55} = 1{,}82$; $\Delta\theta$=55sm $\Delta\xi$=100cm S=60cm, R=66,2 cm

P=60kg=0,6kN, $c^2 = \frac{h^2}{12R^2} = \frac{0{,}25^2}{12\cdot 66{,}2^2} = 0{,}0000011$

$$D = \frac{Eh^3}{12(1-\nu^2)} = \frac{2100000\cdot 0{,}25^3}{12(1-0{,}3)^2} = 300{,}48 kg\cdot cm = 30{,}048 k\ N\cdot cm$$

$$Z_i = q = \frac{P}{\Delta\xi\Delta\theta} = \frac{60}{100\cdot 55} = 0{,}011 kg/cm^2 = 0{,}00011\ kN/cm^2$$

$$Y_i = H = \frac{50{,}45\cdot 0{,}5}{0{,}5(100\cdot 55)}\cos 52^0 = 0{,}5\cdot 0{,}018345\cdot 0{,}62 = 0{,}0113\ kg/cm^2\cdot 0{,}5 = 0{,}006\ kg/sm^2$$

Coefficients:

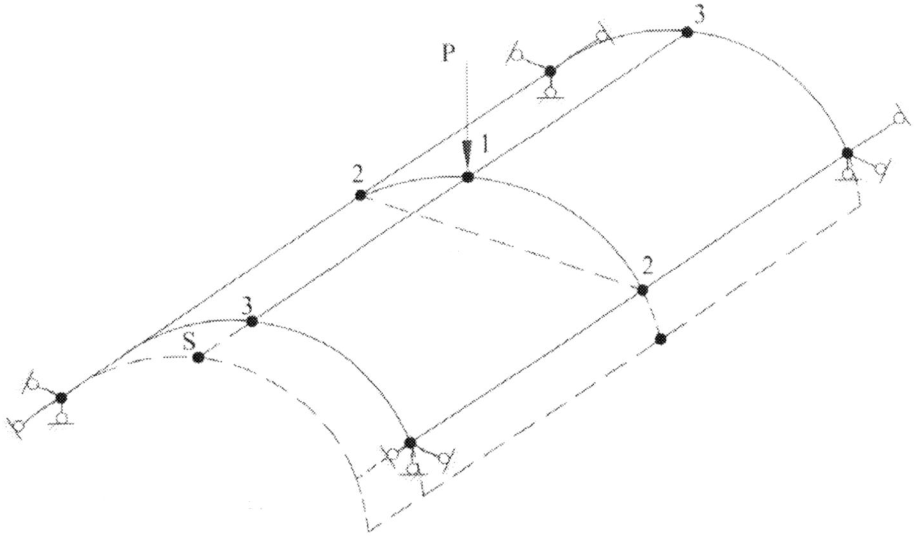

Figure 5.5. Net of test model of the shell.

$$A_\gamma = \frac{1-\nu}{2} = \frac{1-0,3}{2} = 0,35;\; B_\gamma = \frac{1+\nu}{2} = \frac{1+0,3}{2} = 0,65;\; C_\gamma = \frac{3-\nu}{2} = \frac{3-0,3}{2} = 1,35;$$

If we confine only to vertical displacements we shall get algebraic equation system:

$$\left.\begin{array}{l}
27,479\omega_1 - 0,0063\omega_2 - 0,0019\omega_3 + 0\cdot\omega_\upsilon + 0\cdot\omega_s = -0,4596 \\
0\cdot\omega_1 + 54,9525\omega_2 + 0,0029\omega_3 - 0,0063\omega_\upsilon + 0\cdot\omega_s = 0,0120 \\
-0,0019\omega_1 + 0,0029\omega_2 + 54,9584\omega_3 - 0\cdot\omega_\upsilon - 0,0019\omega_s = 0 \\
0,0003\omega_1 + 0\cdot\omega_2 + 10,9627\omega_3 + 0\cdot\omega_v + 0,0003\omega_s = 0 \\
1,0040\omega_1 + 10,8368\omega_2 + 0\cdot\omega_3 + 0,0036\omega_v + 0,0003\omega_s = 0
\end{array}\right\}$$

Algebraic equations system is solved using program "MATCAD"

$$A := \begin{pmatrix} 27,4793 & -0,0063 & -0,0019 & 0 & 0 \\ 0 & 54,9525 & 0,0029 & -0,0063 & 0 \\ -0,0019 & 0,0029 & 54,9584 & 0 & -0,0019 \\ 0,0003 & 0 & 10,9627 & 0 & 0,0003 \\ 1,004 & 10,8368 & 0 & 0,0036 & 0,0003 \end{pmatrix} \quad B := \begin{pmatrix} -0,4596 \\ 0,012 \\ 0 \\ 0 \\ 0 \end{pmatrix}$$

$$\text{lsolve}(A,B) = \begin{pmatrix} -0,017 \\ 5,598\times10^{-4} \\ -1,305\times10^{-8} \\ 2,978 \\ 0,017 \end{pmatrix}$$

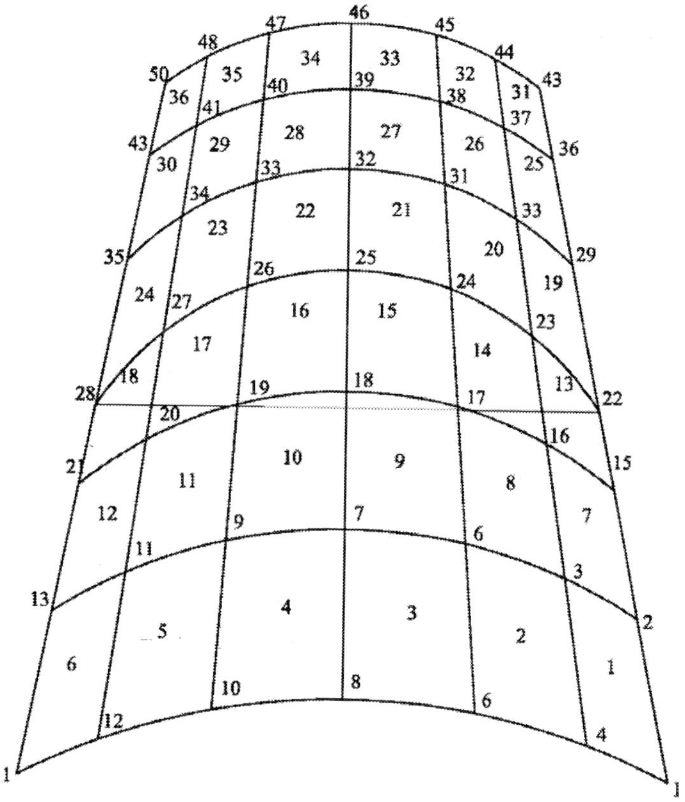

Figure 5.6. Division of shell surface into elements.

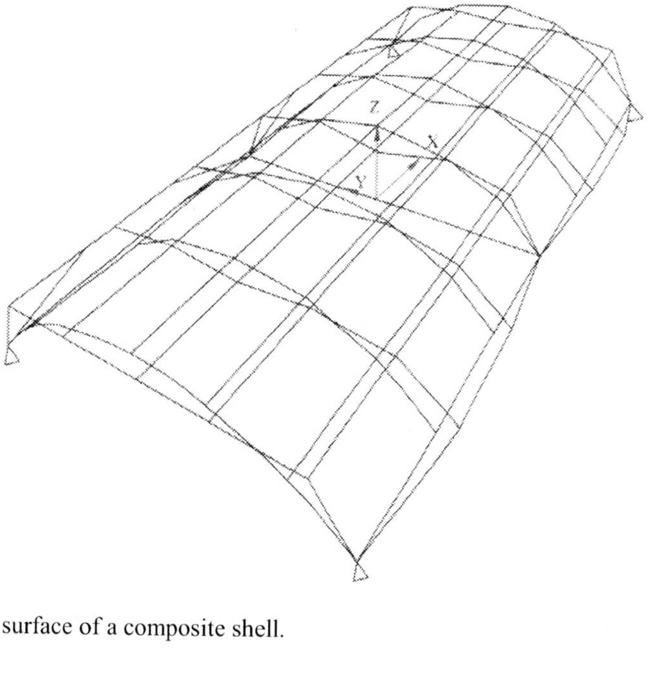

Figure 5.7. Stressed surface of a composite shell.

Figure 5. Experimental model of a cylindrical shell.

Using the parameters of experimental model of cylindrical shell static and dynamic calculations were performed in program "SAP-2000 Student".

Theoretical and test results for vertical displacement of shell before and after switching of tie-bar automatic regulation are given in the form of Table.

From the analysis of the Table it is seen that as a result of automatic regulation of a tie-bar the deflection variation in shell can be:

n_{av}=1,9 (according to test data).

Table 5.1.

Point #	Finite-difference method before switching on of AECM, mm	SAP-2000 Student before switching on of AECM, mm	Point deflection test values before switching on of AECM, mm	Deflection test values after switching on of AECM, mm
1	ω_1=-0,17	ω_1=-1,5657	ω_1=-1,531	ω_1=-2,922
2	$\sqrt{\omega_2^2+\upsilon_2^2}=-3,324$	ω_2=-0,0844	ω_2=-1,432 (1,351)	ω_2=-3,572 (3,124)
3	$\sqrt{\omega_3^2+\upsilon_3^2}=-0,0210$	ω_3=-1,2316	-	-

The difference between theoretical and test results is caused by considerable displacement of the shell because of its thin walls which is not considered in linear theory of shells, also, by improper selection of net with finite-difference method, etc.

In spite of the mentioned differences it is obvious that by means of automatic control the regulation of forces and deflections in shells is possible.

5.4. REGULATION OF FREE OSCILLATION PERIOD IN COMPOSITE SPATIAL CONSTRUCTIONS

The changes of free oscillations period in constructions have a great importance for regulation of their amplitude-resonance phenomena. For example, in the case of calculation for seismic action the graphs of period-dynamic coefficients are used in EUROCOD, as well as, in the normative acts of other countries.

At changing self-oscillation period β coefficient and, respectively, seismic force acting on the construction can be regulated.

5.4.1. Regulation of Composite Cylindrical Shell Period

The equation of composite shell vibrations has the following form:

$$D\frac{\partial^4\omega}{\partial x^4}+\frac{Eh}{R^2}\omega+\rho h\frac{\partial^2\omega}{\partial t^2}+2\rho h\mu\frac{\partial^4\omega}{\partial t^2\partial x^2}\mp N\frac{\partial^2\omega}{\partial x^2}+\delta(x-\xi)M\frac{\partial^2\omega}{\partial t^2}=0 \qquad (5.18)$$

Suppose $\omega(x,t)=W(x)\sin\Omega t$ and substitute (5.4.2) into (5.4.1.), we shall have:

$$D\frac{d^4W}{dx^4}\sin\Omega t+\frac{Eh}{R^2}W\sin\Omega t-\rho h\Omega^2W\sin\Omega t-2\rho h\mu\Omega^2\frac{d^2W}{dx^2}\sin\Omega t+$$
$$\mp N\frac{d^2W}{dx^2}\sin\Omega t-\delta(x-\xi)M\Omega^2W\sin\Omega t=0$$

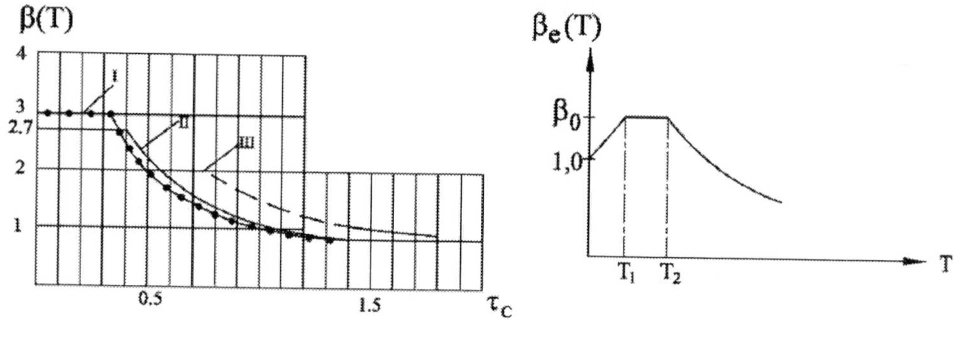

Figure 5.8. Diagrams of dynamic coefficients.

Hence we shall get:

$$D\frac{d^4W}{dx^4}+\frac{Eh}{R^2}W\mp N\frac{d^2W}{dx^2}-\rho h\Omega^2W-2\rho h\mu\Omega^2\frac{d^2W}{dx^2}-M\delta(x-\xi)\Omega^2W=0 \qquad (5.19)$$

Suppose $W=A\sin\frac{n\pi x}{\ell}$ which satisfies boundary value conditions and using Bubnov-Galerkin method, we shall have:

$$\int_0^\ell\left[D\frac{n^4\pi^4}{\ell^4}+\frac{Eh}{R^2}+N\frac{n^2\pi^2}{\ell^2}-\rho h\Omega^2+2\rho h\mu\Omega^2\frac{n^2\pi^2}{\ell^2}-\left[\int_0^\ell\delta(x-\xi)\sin^2\frac{n\pi x}{\ell}\right]M\Omega^2\cdot\right.$$

$$\left.A^2\sin^2\frac{n\pi x}{\ell}dx=0\right.$$

With consideration of local situation of the tie-bar, $N=T^0\delta(x-\eta)$, we shall get:

$$D\frac{n^4\pi^4}{\ell^4}+\frac{Eh}{R^2}\pm T^0\frac{n^2\pi^2}{\ell^2}\frac{2}{\ell}\frac{\ell}{2}\int_0^\ell\delta(x-\Im)\sin^2\frac{n\pi x}{\ell}dx-\rho_\ell h\Omega^2+2\rho h\mu\Omega^2\frac{n^2\pi^2}{\ell^2}$$

$$-M\Omega^2\left(\int_0^\ell\delta(x-\xi)\sin^2\frac{n\pi x}{\ell}dx\right)\ell^3=0$$

If we assume that oscillations are performed without damping μ=0 and the tie-bar is situated in the middle part of shell length $\xi=\eta=\frac{\ell}{2}$ (for square in plan shell) we shall get:

$$\int_0^\ell\delta(x-\Im)\sin^2\frac{2n\pi x}{\ell}dx=\sin^2\frac{n\pi l}{2\ell}=1,0$$

Hereof:

$$\Omega_n^2=\left(D\frac{n^4\pi^4}{\ell^4}+\frac{Eh}{R^2}+T^0\frac{n^2\pi^2}{\ell^2}\right)\left(\frac{1}{\rho h+M}\right) \qquad (5.20)$$

Formula (5.20) in particular case when shell oscillations are performed only in longitudinal direction and M=0 coincide with O.Oniashvili formula $\omega_{mn}^2=\frac{g}{\gamma\delta R^2}\left[\frac{D}{R^2}\left(\lambda_n^2+\mu_m^2\right)^2+\left(T_1^0\lambda_n^2+T_2^0\mu_m^2\right)+\frac{E\delta\lambda_n^4}{\left(\lambda_n^2+\mu_m^2\right)^2}\right]$, formula (2.19). Here

$\lambda_n = \frac{n\pi R}{\ell}$ and $\mu_n = \frac{m\pi}{\beta_0}$; from formulas (5.4.4) and (2.19) we receive that prestretching causes frequency increase, and decreases free oscillations frequency. Here T_1^0 and T_2^0 are compressing or stretching forces on shell linear unit.

If we insert the parameters of test cylindrical shell into the mentioned formula (5.20) we shall get (in case of two half-waves n=2):

$$D = \frac{Eh^3}{12(1-\mu^2)} = \frac{2{,}1\cdot 10^{-6}\cdot 0{,}25^3}{12(1-0{,}3^2)} = 3004{,}8 kgsm, \quad \frac{n^4\pi^4}{\ell^4} = \frac{2^4\cdot 3{,}14^4}{110^4} = 0{,}0000424;$$

$$D\frac{n^4\pi^4}{\ell^4} = 3004{,}8\cdot 0{,}0000424 = 0{,}127 kg/sm^3; \quad \rho = \frac{\gamma}{g} = \frac{0{,}00781}{981} = 0{,}000008\frac{kg\sec^2}{sm^2};$$

$$T^0\frac{n^2\pi^2}{\ell^2} = 0{,}0033\cdot 5 = 0{,}0165\frac{kg}{sm^3}$$

$$\frac{Eh}{R^2} = \frac{2{,}1\cdot 10^6\cdot 0{,}25}{66{,}2^2} = 119{,}79 kg/sm; \quad \rho\text{h}=0{,}25\cdot 0{,}000008=0{,}000002\frac{kg\sec^2}{sm}$$

$$M = \frac{G}{g} = \frac{60}{981} = 0{,}061\frac{kg\sec^2}{sm}; \quad \rho\text{h+M}=0{,}000002+0{,}061=0{,}061002\frac{kg\sec^2}{sm}$$

$$\Omega^2 = (0{,}127+119{,}79+0{,}0165)/0{,}061002 = 1966{,}058 rad/\min^2; \quad \Omega=44{,}34 rad/\sec$$

$$T = \frac{2\pi}{\Omega} = \frac{6{,}28}{44{,}34} = 0{,}1416\sec; \quad \text{SAP-2000 T=0,1022sec}$$

5.4.2. Regulation of Free Vibrations Period of Composite Tent-wise Shell

The equation of vibrations of composite tent-wise shell strengthened with strut system on the basis of declined shells theory will be expressed as (Figure 1):

Figure 1.

$$\frac{1}{Eh_{red}}\nabla^4\Phi - \Gamma^{(1)}_{(\alpha=0)}\theta\frac{\partial w(0,\beta)}{\partial\beta^2} - \Gamma^{(1)}_{(\beta=0)}\theta\frac{\partial w(\alpha,0)}{\partial\alpha^2} = 0$$

$$D\nabla^4 w + \Gamma^{(1)}_{(\alpha=0)}\theta\frac{\partial^2\Phi(0,\beta)}{\partial\beta^2} + \Gamma^{(1)}_{(\beta=0)}\theta\frac{\partial^2\Phi(\alpha,0)}{\partial\alpha^2} + \frac{\gamma h_{red}}{g}\frac{\partial^2 w}{\partial t^2} \pm \tag{5.21}$$

$$\pm N_\alpha\frac{\partial^2 w}{\partial\alpha^2} \pm N_\beta\frac{\partial^2 w}{\partial\beta^2} - k_0(\alpha,\beta)w(\alpha,\beta) = 0$$

Here $K_0(\alpha,\beta) = K_0\delta\left(\alpha - \frac{a}{2}\right)\cdot\delta\left(\beta - \frac{\beta}{2}\right)$

θ is the angle between two adjacent faces;

K_0 is strut system flexibility coefficient taken in roof centre;

h_{red} is the reduced thickness of the shell;

Here and below the following notations are used:

a, b, α_0 and β_0 are roofing dimensions in plan;

N_α, N_β, N_{0x} and N_{0y} are longitudinal horizontal forces on roofing contour transmitted from strut system;

δ is Dirack function;

θ is roof surface refractive angle;

m, n, μ and i are wave numbers;

γ is roof material volumetric weight;

G is gravity acceleration;

R is radius;

D is cylindrical rigidity of shell $D = \dfrac{Eh_{red}}{12(1-\nu^2)}$;

E is elasticity modulus;

γ is Poisson coefficient;

ω is free vibrations frequency;

T is period, $T = \dfrac{2\pi}{\omega}$ sec.

Express $\Phi(\alpha, \beta, t)$ function of stress and deflection on tent-wise shell profile in case of joint fastening as:

$$\Phi(\alpha,\beta,t) = \sin\Omega t \sum_{m=1}^{\infty}\sum_{n=1}^{\infty} A_{mn} \sin\frac{m\pi\alpha}{a}\cdot\sin\frac{m\pi\beta}{b}$$
$$\Phi(\alpha,\beta,t) = \sin\Omega t \sum_{m=1}^{\infty}\sum_{n=1}^{\infty} B_{mn} \sin\frac{m\pi\alpha}{a}\cdot\sin\frac{m\pi\beta}{b} \tag{5.22}$$

In this expression the following substitution is to be done:

$$y = \frac{\sqrt{2}}{2}(\alpha+\beta) \text{ and } x = \frac{\sqrt{2}}{2}(\alpha-\beta) \tag{5.23}$$

If we use Bubnov-Galerkin method we shall get the frequency of tent-wise shell free oscillation (main tone) strengthened with a strut system:

$$\omega_{main}^2 = \frac{g}{vh_{red}}\frac{D\pi^4}{4a^4} + \frac{\theta^2 Eg}{3,56a^2\gamma} + \frac{k_0}{2a^2}\cdot\frac{g}{\gamma h_{red}} \mp \frac{\pi^2 g}{4a^2\gamma h_{red}}(N_\alpha + N_\beta) \tag{5.24}$$

5.4.3. Regulation of Self Vibration Period of a Slab Strengthened with Guy System

Consider a thin walled slab which is jointly supported on contour and strengthened with intercrossing diagonal guy system (Figure 2).

The equation of slab movement has the following form:

$$\frac{D}{h_{red}}\nabla^2\nabla^2 w = L(w,\Phi) - \frac{\gamma}{g}\frac{\partial^2 w}{\partial t^2} \mp N_x\delta\left(x-\frac{a}{2}\right)\frac{\partial^2 w}{\partial x^2} \mp N_y\delta\left(y-\frac{b}{2}\right)\frac{\partial^2 w}{\partial y^2} - \frac{k_0}{h_{red}}\delta\left(x-\frac{a}{2}\right)\left(y-\frac{b}{2}\right)w;$$
$$\frac{1}{E}\nabla^4\Phi = -\frac{1}{2}L(w,w) \tag{5.25}$$

where

$$L(w,\Phi) = \frac{\partial^2 w}{\partial x^2}\cdot\frac{\partial^2 \Phi}{\partial y^2} + \frac{\partial^2 \Phi}{\partial x^2}\cdot\frac{\partial^2 w}{\partial y^2} - 2\frac{\partial^2 w}{\partial x\partial y}\cdot\frac{\partial^2 \Phi}{\partial x\partial y};$$
$$L(w,W) = 2\left[\frac{\partial^2 w}{\partial x^2}\cdot\frac{\partial^2 w}{\partial y^2} - \left(\frac{\partial^2 w}{\partial x\partial y}\right)^2\right]; \tag{5.26}$$

Stress function in the case of compressed forces represent as:

$$\Phi(x,y,t) = \frac{ET(t)^2}{32}\left[\left(\frac{a}{b}\right)^2\cos 2\frac{\pi x}{a} + \left(\frac{b}{a}\right)^2\cos 2\frac{\pi y}{b}\right] - \frac{\overline{P}_x\cdot y^2}{2} - \frac{\overline{P}_y\cdot x^2}{2} \tag{5.27}$$

Contour forces are determined with expression:

$$\overline{P}_x = E\frac{\pi^2}{8b^2}\cdot\frac{\left[\nu + \frac{b^2(\nu_y - 1)}{a^2}\right]}{\left[(\nu_x-1)(\nu_y-1)-\nu^2\right]}T^2(t)$$
$$\overline{P}_y = E\frac{\pi^2}{8a^2}\cdot\frac{\left[\nu + \frac{a^2(\nu_y - 1)}{b^2}\right]}{\left[(\nu_x-1)(\nu_y-1)-\nu^2\right]}T^2(t) \tag{5.28}$$

where $\nu_x = \frac{h_{red}}{F_x}$ and $\nu_y = \frac{h_{red}}{F_y}$; here F_x and F_y are the area of slab profile edges per unit length (A.C.Volmir, Nonlinear dynamics of plates and shells. Ed. Nauka, 1972, p.76, 432 p).

Suppose the solution of equation (5.25) as:

$$w(x,y,t) = T(t)\sin\frac{i\pi x}{a}\sin\frac{i\pi y}{b} \tag{5.29}$$

which satisfies the boundary condition and using Bubnov-Galerkin method we shall get main frequency of free vibrations of a strengthened slab:

$$\omega^2 = \frac{D}{h_{red}}\frac{g\pi^4}{\gamma}\left(\frac{1}{a^2}+\frac{1}{b^2}\right)^2 + \frac{4g}{\gamma}\frac{k_0}{abh_{red}} \pm \frac{2\pi^2 g}{\gamma}\left(\frac{N_{0x}}{a^3}+\frac{N_{0y}}{b^3}\right) \tag{5.30}$$

5.4.4. Regulation of Free Vibration Period of a Composite Equilateral Shell

Vibration equation of a composite equilateral shell strengthened with a strut system (Figure 3) is expressed as:

$$\frac{1}{Eh_{red}}\nabla^4\Phi(\alpha_1,\beta) - \frac{1}{R}\frac{\partial^2 w(\alpha_1,\beta)}{\partial\alpha^2} - \Gamma^{(1)}_{\left(\beta=\beta_1=\frac{\beta_0}{2}\right)}\theta\cdot\frac{\partial^2 w(\alpha_1,\beta_1)}{\partial\alpha^2} = 0$$

$$D\nabla^4 w(\alpha_1,\beta) + \frac{1}{R}\frac{\partial^2\Phi(\alpha_1,\beta)}{\partial\alpha^2} + \Gamma^{(1)}_{\left(\beta=\beta_1=\frac{\beta_0}{2}\right)}\theta\cdot\frac{\partial^2\Phi(\alpha_1,\beta_1)}{\partial\alpha^2} + \frac{\gamma h_{red}}{g}\frac{\partial^2 w}{\partial t^2} +$$

$$+N_\alpha\frac{\partial^2 w}{\partial\alpha^2} + N_\beta\frac{\partial^2 w}{\partial\beta^2} + k_0(\alpha_1,\beta)w(\alpha_1,\beta) = 0$$

where $k_0(\alpha_1,\beta) = k_0\delta\left(\alpha_1 - \frac{\alpha_0}{2}\right)\delta\left(\beta_1 - \frac{\beta_0}{2}\right)$ (5.31)

$\Gamma^{(1)}$ is unit pulse function (O.D.Oniashvili, Some dynamic problems of the theory of shells, Ed. AS USSR, 1957, 194p).

Consider equilateral shell joint support and obtain the following:

$$\Phi(\alpha_1,\beta,t) = \sin\Omega t\sum_{n=1}^{\infty}\sum_{m=1}^{\infty}A_{n,m}\sin\frac{n\pi\alpha}{\alpha_0}\sin\frac{m\pi\beta}{\beta_0}$$

$$W(\alpha_1,\beta,t) = \sin\Omega t\sum_{n=1}^{\infty}\sum_{m=1}^{\infty}B_{n,m}\sin\frac{n\pi\alpha}{\alpha_0}\sin\frac{m\pi\beta}{\beta_0} \tag{5.32}$$

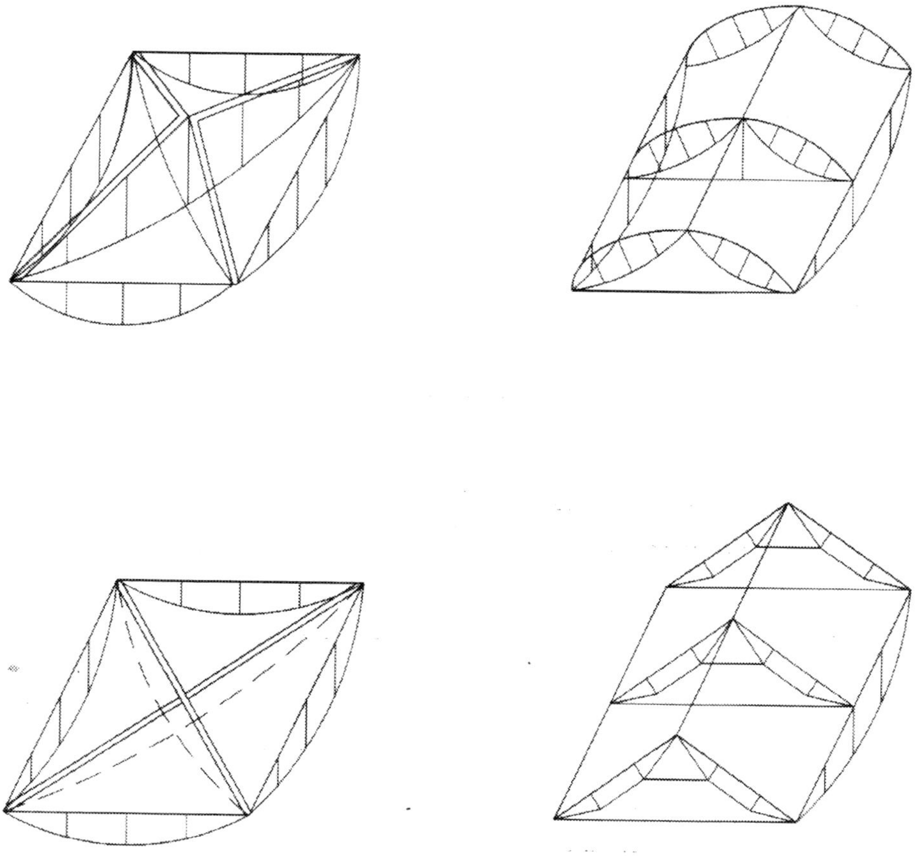

Figure 5.9. Types of composite shells.

Using Bubnov-Galerkin method and considering the properties of pulse $\Gamma^{'}$ and δ functions we shall have:

$$\omega_{n,m}^{2} = \frac{g}{\gamma h_{red}} \Big[D\left(\lambda^{*2} + \mu^{*2}\right)^{2} + \frac{E h_{red} \lambda^{*4} \left(\frac{1}{R} + \frac{2\theta}{\beta_0}\right)^{2}}{\left(\chi^{*2} + \mu^{*2}\right)^{2}} + k_0 \frac{4}{\alpha_0 \beta_0} \mp \pi^{2} \left(N_{\alpha} \frac{n^{2}}{\alpha_0^{2}} + N_{\beta} \frac{m^{2}}{\beta_0^{2}} \right) \qquad (5.33)$$

where $\lambda^{*} = \frac{n\pi}{\alpha_0}$ and $\mu^{*} = \frac{m\pi}{\beta_0}$.

5.4.5. Regulation of the Period of Self Vibrations of Composite Folded-plate Shell

The equation of self vibrations of composite folded-plate shell strengthened with guy system (Figure 4) is expressed as:

$$\frac{1}{Eh}\nabla^4\Phi - \Gamma^{(1)}_{(\beta_1)}\theta\frac{\partial^2 w(\alpha,\beta_1)}{\partial\alpha^2} = 0$$

$$D\nabla^4\Phi - \Gamma^{(1)}_{(\beta_1)}\theta\frac{\partial^2\Phi(\alpha,\beta_1)}{\partial\alpha^2} + \nu\frac{h_{red}}{g}\frac{\partial^2 w}{\partial t^2} \pm N_\alpha\frac{\partial^2 w}{\partial\alpha^2} \pm N_\beta\frac{\partial^2 w}{\partial\beta^2} - \qquad (5.34)$$

$$-k_0(\alpha,\beta)w(\alpha,\beta) = 0$$

Let's the functions of voltage $\Phi(\alpha,\beta)$ and bending $w(\alpha,\beta)$ be:

$$\Phi(\alpha,\beta) = \sin\Omega t\sum_{m=1}^{\infty}\sum_{n=1}^{\infty}A_{mn}\sin\frac{m\pi\alpha}{a}\sin\frac{n\pi\beta}{b}$$

$$W(\alpha,\beta) = \sin\Omega t\sum_{m=1}^{\infty}\sum_{n=1}^{\infty}B_{mn}\sin\frac{m\pi\alpha}{a}\sin\frac{n\pi\beta}{b} \qquad (5.35)$$

If we use Bubnov-Galerkin method we shall receive the formula of free vibrations frequency for composite folded-plate shell:

$$\omega^2_{n,m} = \frac{g}{\gamma h_{red}}\left[D\pi^4\left(\frac{m^2}{a^2}+\frac{n^2}{b^2}\right) + k_0\frac{4}{a\cdot b} + \frac{4\theta^2 m^4 Eh}{a^4 b^2\left(\frac{m^2}{a^2}+\frac{n^2}{b^2}\right)} \pm \left(N_\alpha\frac{m^2}{a^2} + N_\beta\frac{n^2}{b^2}\right)\pi^2\right]$$

(5.36)

The forms of folded-plate shell vibrations have nodal lines and swellings that may coincide with fold edge, as well as, with guy support; as a result their effect on vibrations frequency is decreased.

5.5. Numerical Realization of Regulation of Free Vibrations Period for Composite Spatial Structures

From the presented composite spatial structures two types of shells - tent-like and folded-plate ones were chosen as numerical example.

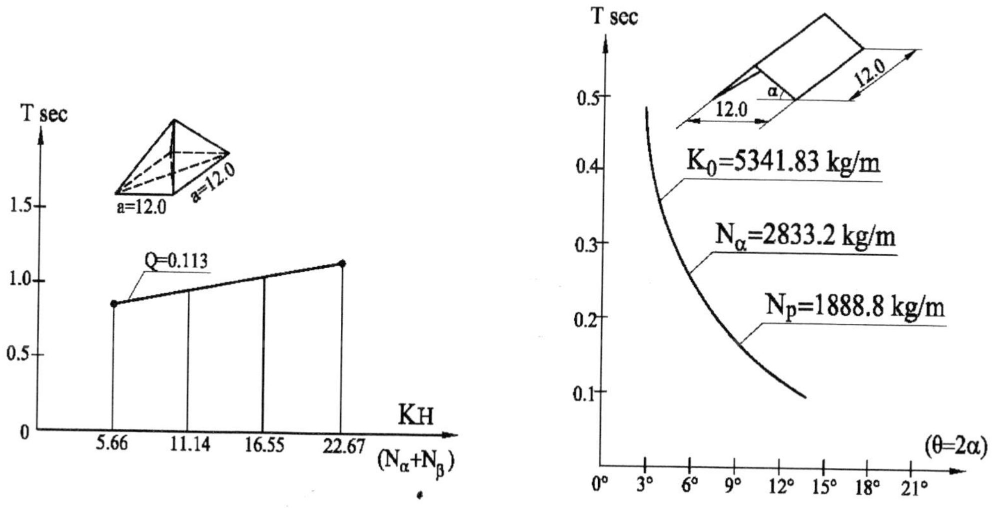

Figure 5.10. The relation between cable tension and shell angle variation and period.

Structural material for the chosen shells was glass textolyte of KACT-B brand with elasticity module E= 44 400 kg/cm and Poisson coefficient $\nu = 0.49$; volume weight $\gamma = 1.4$ t/m; in cable system ∅32 mm cable was used; shell dimensions are a=b=12.0 m; reduced thickness of shell h_{red}=5.0cm; cylindrical rigidity of shell $D = Eh_{red}^3/12(1-\nu^2) = 60.8633$kg/m; angle of fold $\theta=2\alpha=0.113$; cable strain changes from 17 tons to 68 tons; yielding k_0 of the middle support of tent-like shell was respectively changed.

As a result of the carried out calculations with cable strain regulation it is possible to change period variation for 24.16%. As is seen from the Figure the mentioned relation is linear in the case of folded-plate shell. Maintaining permanent strain of cables we regulate the fold inclination angle θ and receive the variation of self oscillation period for 80.34%. Here the dimensions of folded-plate shell were a=12 m and b=18.0; the other characteristics are similar to those of tent-like shells. As is seen from the Figure the dependence is nonlinear.

Chapter 6

REGULATION OF VIBRATIONS OF SUSPENSION AND GUY BRIDGES USING ELECTRO-MECHANICAL AND FIBER-OPTICAL STRUCTURES

6.1. INTRODUCTION

Bridges in which cables and guys are used as support elements are of two types: suspension bridges and guy bridges (Figure 6 a, b, and c).

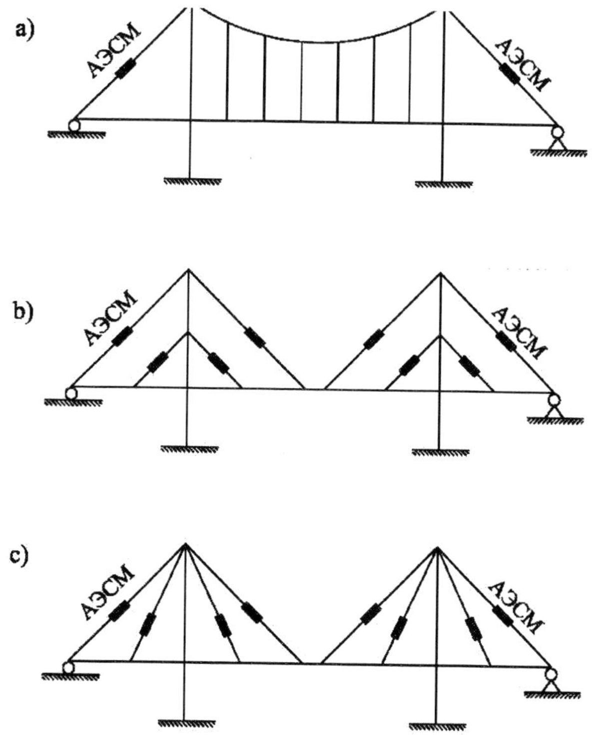

Figure 6.1. Diagrams of suspension and guy bridges with automatic tie-coupling on cables.

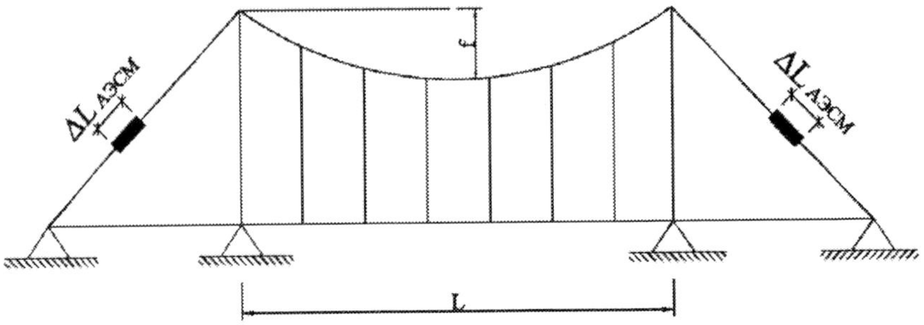

Figure 6.2. Regulation of vibration freguency of hanging bridge.

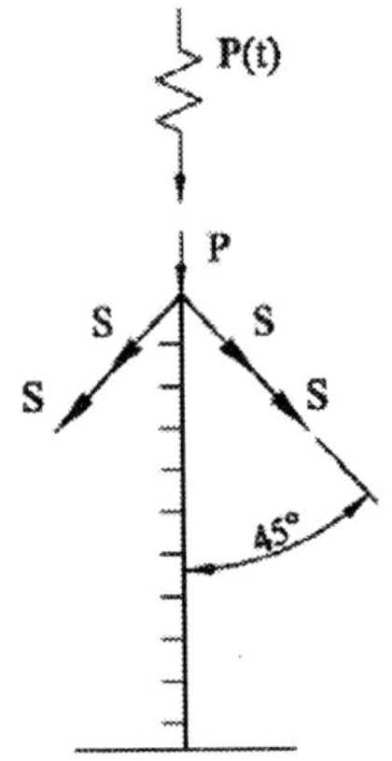

Figure 6.3. Regulation of aerodynamic stability of suspension bridge pylon.

6.1. Regulation of Vibrations of Suspension Bridges

Assume the equation of hanging bridge movement in the following form:

$$EI\frac{\partial^4 y}{\partial x^4}+(m_0+m)-H(x_1t)\frac{\partial^2 y}{\partial x^2} \tag{6.1}$$

The solution of equation (6.1) express in the form:

$$y = X\sin\frac{n\pi x}{\ell} \tag{6.2}$$

and use Bubnov-Galerkin method:

$$\frac{\partial y}{\partial x}=X\frac{n\pi}{\ell}\cos\frac{n\pi x}{\ell};\ \frac{\partial^2 y}{\partial x^2}=-X\frac{n^2\pi^2}{\ell^2}\sin\frac{n\pi x}{\ell}$$

$$\frac{\partial^2 y}{\partial t^2}=\frac{\partial^2 x}{\partial t^2}\sin\frac{n\pi x}{\ell}$$

$$\mathrm{P(x,t)=P_1}(t)\sin\frac{n\pi x}{\ell}=P_0\cos\theta t\sin\frac{n\pi x}{\ell} \tag{6.3}$$

$$\frac{\partial^4 y}{\partial t^4}=X\frac{n^4\pi^4}{\ell^4}\sin\frac{n\pi x}{\ell}$$

If we introduce equation (6.3) into equation (6.1) we'll get

$$E\Im\cdot\frac{n^4\pi^4}{\ell^4}\int_0^\ell\sin^2\frac{n\pi x}{\ell}dx+(m_0+m)\frac{\partial^2 x}{\partial t^2}\int_0^\ell\sin^2\frac{n\pi x}{\ell}dx+X\frac{n^2\pi^2}{\ell^2}\int_0^\ell H(x,t)\sin^2\frac{n\pi x}{\ell}dx=$$
$$=P_0\cos\theta t\int_0^\ell\sin^2\frac{n\pi x}{\ell}dx \tag{6.4}$$

Here m_0 and m are constant and temporary equidistributed mass per one linear meter of bridge;

$E\Im$ is the rigidity of sliding beam;

H(x,t) - static and dynamic loads in bunton cable is $H(x,t)=H_0+\delta(x)H_1$;

$\delta(x)$ is Dirack function;

ℓ is the span of hanging bridge.

After integrating (6.4) we get:

$$E\Im\cdot x\frac{n^4\pi^4}{\ell^4}\frac{\ell}{2}+(m_0+m)\frac{d^2x}{dt^2}\frac{\ell}{2}+x\frac{n^2\pi^2}{\ell^2}\left(H_0\frac{\ell}{2}+H_1\sin^2\frac{n\pi x}{\ell}\right)=P_0\cos\theta t\frac{\ell}{2} \tag{6.5}$$

Dividing equation (6.5) by l/2 we get:

$$E\Im\cdot x\frac{n^4\pi^4}{\ell^4}+(m_0+m)\frac{d^2x}{dt^2}+x\frac{n^2\pi^2}{\ell^2}H_0+\frac{n^2\pi^2}{\ell^2}H_1\frac{2}{\ell}\cdot\sin^2\frac{n\pi a}{\ell}=P_0\cos\theta t \tag{6.6}$$

or

$$\frac{\partial^2 x}{\partial t^2}+X\left[\frac{n^4\pi^4\cdot E\Im}{\ell^4(m_0+m)}+\frac{n^2\pi^2}{\ell^2}\frac{H_0}{(m_0+m)}+\frac{2n^2\pi^2}{\ell^3}\frac{H_1}{(m_0+m)}\sin^2\frac{n\pi a}{\ell}\right]=$$
$$=\frac{P_0}{(m_0+m)}\cos\theta t \tag{6.7}$$

Denote

$$\omega^2 = \frac{n^4\pi^4 \cdot E\Im}{\ell^4 (m_0 + m)} + \frac{n^2\pi^2}{\ell^2} \frac{H_0}{(m_0 + m)} + \frac{2n^2\pi^2}{\ell^3} \frac{H_1}{(m_0 + m)} \sin^2 \frac{n\pi a}{\ell} \tag{6.8}$$

$$G = \frac{P_0}{(m_0 + m)} \tag{6.9}$$

Equation (6.7) is expressed as:

$$\frac{d^2x}{dt^2} + \omega^2 x = G\cos\theta t \tag{6.10}$$

Find the solution of (6.10) in the following form X=Acosθt, we shall get: $\frac{d^2x}{dt^2} = -A\theta^2 \cos\theta t$ and introduce in (6.10):

$$-A\theta^2 \cos\theta t + \omega^2 A\cos\theta t = G\cos\theta t \tag{6.11}$$

reducing for cos θt

we shall get: $-A\theta^2 + \omega^2 A = G$ (6.12)

The amplitude of induce vibrations is:

$$A = G/\left(\omega^2 - \theta^2\right) \tag{6.13}$$

Dynamic bunton in suspension bridge cable can be expressed with the following approximate formula:

$$H = \frac{16 f_b E_b F_n A}{3\ell^3 \left[1 + \frac{16}{3}\left(\frac{f_z}{\ell}\right)^2\right]} \quad \text{when n=1,3,5...} \tag{6.14}$$

Here n is the number of vibrations modes;

A is vibration amplitude;

$f\ E\ F$ are the sag of hanging bridge cable, cable flexibility module and cross-section area.

θ is the frequency of induced vibrations; total bunton: $H=H_{st}+H_{din}$

Gap magnitude $\Delta\ell_{AECM}$ between AECM contactors according to Hook law $\sigma = \varepsilon E$ is:

$$\frac{H}{F_{red}} = \frac{\Delta\ell_{АЭСМ}}{\ell_{red}} \cdot E_{red} \tag{6.15}$$

Here of:

$$\Delta\ell_{АЭСМ} = \frac{H\ell_{red}}{E_{red}F_{red}} = \frac{\ell_{red}}{E_{red}F_{red}}\left[\frac{16 f_b E_b F_b A}{3\ell^2\left[1+\frac{16}{3}\left(\frac{f_b}{\ell}\right)^2\right]} + \frac{\int_0^l \frac{M_1 M_2}{EI}dx}{\int_0^l \frac{M_2^2 dx}{EI} + \sum N^2 \frac{S}{EF}}\right] \tag{6.16}$$

$$\text{Here } A = \frac{P_0}{(m_0+m)(\omega^2-\theta^2)} \tag{6.17}$$

Aerodynamic Stability of Cable-braced Bridge Pylon

The construction of pylons is quite varied. This mainly depends on the used material (stone, steel, reinforced concrete), bridge system, architectural solution, etc.

Cable-bridge pylons are mainly of two kinds: flexible and rigid.

Flexible pylons are single support and they can less have endurance to horizontal forces. For small spans on their origin point the cables are fixed jointly for the simplification of mounting.

Rigid pylons are fixed so that they are not horizontally displaced even as a result of pylon work.

We discuss one support flexible pylon which is rigidly fixed with one end in support abutment or in rigid beam, on the second end the joint origin is crossed with guy which transmits vertical static and dynamic loads to the pylon (Figure 6.3). Pylon body is made of cylindrical steel pipe.

Wind flow simultaneously acts on pylon body and guys and at a definite velocity causes the vibration of guys, as well as, pylon body.

The purpose of the work is to determine critical velocity of wind on guy bridge pylon when pylon losses aerodynamic stability (see Figure 6.3).

Determination of Loads

The loads on guy bridge pylon are determined by static and dynamic action of wind and with static and dynamic forces on it induced in cables.

Static load of wind on cylindrical elements is determined with formula:

$$q = \frac{1}{2}\rho v^2 d C_{L_0} n$$

where ρ is air density;

v is wind velocity;

d is the diameter of cylindrical element;

C_{L0} is aerodynamic coefficient.

The loading from guy bridge roadway and the component of T force induced in guy because of wind load is transferred as horizontal force to pylon and is equal to P_l=2Tsinα, where angle α between guy and pylon is analogously equal to $P_0 = 2T_{0}\sin\alpha$ where $T_0 = 0,445\sqrt[3]{q^2 l^2 E}$, guy frequency is determined with formula:

$$\Omega = \frac{n\pi}{l}\sqrt{\frac{T+T_0}{m}}$$

where l is guy span;

m is linear meter mass of a of guy

Aerodynamic Stability of One Support Pylon

One support cylindrical pylon of a guy bridge is considered on which pulsing wind flow, as well as, horizontal dynamic loading of guys fixed on pylon top are acting. Thus, if we present the pylon as a cylindrical shell, the equation of its motion will have the form:

$$\frac{\partial^4\omega}{\partial x^4} + \xi_0\frac{\partial^2\omega}{\partial t} + \frac{r^2}{\overline{\overline{D}}} = P(t)\frac{\partial^2\omega}{\partial x^2} + \frac{r(1-y^2)}{Ea^{-2}}\rho(t)\frac{\partial^2\omega}{\partial t^2} = \frac{r^4}{\overline{\overline{D}}}q(x,t) \tag{6.18}$$

where ω is radial displacement of cylindrical shell;

$\overline{\overline{D}}$ bending rigidity =E1

y and E are Poisson coefficient and module of elasticity;

r is shell radius; $a^{-2} = \frac{h^2}{12r^2}$

h is shell thickness;

ρ is material density;

l is pylon height.

Assume $\omega(x,t) = W(x)T(t)$, introducing notations and using Bubnov-Galerkin method we receive:

$$\ddot{T} + \mu_1\dot{T} + \omega^2(1 - K\cos 2\Omega t)T = b\cdot e\frac{iVSh}{d}t$$

where

$$\mu_1=\frac{R_2}{R_5};\quad k=\frac{P_1\dfrac{R_3}{R_5}}{\dfrac{R_1}{R_5}+P_0\dfrac{R_3}{R_5}};\quad P(t)=P_0-P_1\cos 2\Omega t$$

$$b=\frac{R_4}{R_5};\quad \omega^2=\frac{R_1}{R_5}+P_0\frac{R_3}{R_5}$$

Here:

$$R_1=\int_0^l\frac{d^4W(x)}{dx^4}W(x)dx;\quad R_2=\mu_0\int_0^l W^2(x)dx;\quad R_3=\frac{r^2}{\overline{\overline{D}}}\int_0^l\frac{d^2W(x)}{dx^2}W(x)dx;$$

$$R_4=\frac{r^4}{\overline{\overline{D}}}\cdot\frac{1}{2}\rho_V\cdot V^2d\cdot C_{L_0}\int_0^l W\ (x)dx;\quad R_5=\frac{r\left(1-\nu^2\right)}{Ea^{-2}}\rho\int_0^l W^2(x)dx.$$

where V is wind velocity;

Sh is Strouhall number;

$i=\sqrt{-1}$;

Ω is guy vibration frequency;

ρ is wind flow density;

C_{Lc} is aerodynamic coefficient;

d is pylon diameter.

Introduce new variable:

$$\tau=\Omega t;\quad T(t)=y(\tau)e^{\frac{-\mu\tau}{2}};\quad \alpha=\frac{\mu_1}{2}$$

The expression of relation between the derivatives of old argument and function t and new argument T(t) and function y(τ) is:

$$\frac{dy}{dx}=\frac{\dfrac{\partial\psi}{\partial t}+\dfrac{\partial\psi}{\partial u}\cdot\dfrac{du}{dt}}{\dfrac{\partial\varphi}{\partial t}+\dfrac{\partial\varphi}{\partial u}\cdot\dfrac{du}{dt}};$$

$$\frac{d^2y}{dx^2}=\frac{1}{\dfrac{\partial\varphi}{\partial t}+\dfrac{\partial\varphi}{\partial u}\cdot\dfrac{du}{dt}}\cdot\frac{d}{dt}\left[\frac{\dfrac{\partial\psi}{\partial t}+\dfrac{\partial\psi}{\partial u}\cdot\dfrac{du}{dt}}{\dfrac{\partial\varphi}{\partial t}+\dfrac{\partial\varphi}{\partial u}\cdot\dfrac{du}{dt}}\right]$$

In our case:

$$\frac{dT(t)}{dt}=\frac{\Omega\left[\psi'(\tau)-\alpha\psi(\tau)\right]}{e^{\alpha t}};$$

$$\frac{d^2T(t)}{dt^2}=\frac{\Omega^2\left[\psi''(\tau)-2\psi'(\tau)\cdot\alpha-\psi(\tau)\alpha^2\right]}{e^{\alpha t}};$$

If we insert (6.18) in the equation and suppose that attenuation $\mu_l=0$ and respectively,α = 0, we shall receive Mathieu type inhomogeneous equation:

$$\frac{d^2y(t)}{d\tau^2}+\left(a-2q\cos 2\tau\right)y(\tau)=f(\tau) \tag{6.19}$$

Here

$$\psi(\tau)=y(\tau)\quad and\quad f(\tau)=r\cdot e^{ip\tau};\ \ p=\frac{\Theta}{\Omega};\ \ a=\left(\frac{\omega}{\Omega}\right)^2;\ \ 2q=k\left(\frac{\omega}{\Omega}\right)^2$$

Let's consider linear inhomogeneous differential equation with periodical coefficients:

$$\frac{d^2u}{dx^2}+p(x)\frac{du}{dx}+q(x)u=f(x) \tag{6.20}$$

Assume that $u=u_l$ is the particular solution of this equation, so that:

$$u_1''+p(x)u_1'+q(x)u_1=f(x) \tag{6.21}$$

Introducing a new function y instead of u we shall receive $u=y+u_l$
Insert into equation (6.20):

$$\left[y''+p(x)y'+q(x)y\right]+\left[u_1''+p(x)u_1'+q(x)u_1\right]=f(x)$$

From the condition of equation (6.21) we shall get:

$$y''+p(x)y'+q(x)y=0 \tag{6.22}$$

We received linear homogeneous differential equation with periodic coefficients which has two linear independent solutions
y_1 and y_2,
then: $u=c_1y_1+c_2y_2+u_l$

Assume that constants C_1 and C_2 are the functions of x so that they are $v_1(x)$ and $v_2(x)$, then if the solution of homogeneous equation y_1 and y_2 is known the general solution will be:

$$y = c_1y_1 + c_2y_2$$

If we represent y_1 and y_2 in the following form:

$$y_1 = e^{iax} \text{ and } y_2 = e^{-iax}$$

then $y_1 = c_1e^{iax} + c_2e^{-iax}$

where $c_1 = \int \varphi_1(x)dx$ *ande* $c_2 = \int \varphi_2(x)dx$

The general solution will be:

$$y = e^{iax} \int \varphi_1(x)dx + e^{-iax} \int \varphi_2(x)dx$$

We receive equation system:

$$\begin{cases} c_1'y_1 + c_2'y_2 = 0 \\ c_1'y_1' + c_2'y_2' = f(x) \end{cases}$$

Vronski determinant:

$$\delta = \begin{vmatrix} y_1 & y_2 \\ y_1' & y_2' \end{vmatrix} = y_1y_2' - y_2y_1' \neq 0$$

$$\delta_1 = \begin{vmatrix} 0 & y_2 \\ f(x) & y_2' \end{vmatrix} = 0y_2' - y_2f(x) = y_2f(x)$$

$$\delta_2 = \begin{vmatrix} y_1 & 0 \\ y_1' & f(x) \end{vmatrix} = y_1f(x) - 0y_1' = y_1f(x)$$

Hereof:

$$\varphi_1(x) = \dot{c}_1 = \frac{\delta_1}{\delta} = \frac{-y_2f(x)}{\delta};$$

$$\varphi_2(x) = \dot{c}_2 = \frac{\delta_2}{\delta} = \frac{y_1f(x)}{\delta};$$

or $y = e^{Lax} \int -\frac{y_2 f(x)}{\delta} dx + e^{-Lax} \int -\frac{y_1 f(x)}{\delta} dx + \text{particular solution}$

$$\text{or } y = -x_1(\tau) \int_0^t \frac{x_2(\tau) t(\tau)}{\delta} d\tau + x_2(\tau) \int_0^t \frac{x_1(\tau) t(\tau)}{\delta} d\tau + \text{ particular solution} \quad (6.23)$$

where $x_1(\tau)$ and $x_2(\tau)$ are particular solutions of Mathieu equation.

According to Floquet theorem partial solutions assume $x_1(\tau)$ and $x_2(\tau)$ in the form:

$$x_1(\tau) = e^{i\beta\tau} \sum_{r=-\infty}^{\infty} c_{2r} \cdot e^{i2r\tau}$$

$$x_2(\tau) = e^{-i\beta\tau} \sum_{r=-\infty}^{\infty} c_{2r} \cdot e^{-i2r\tau}$$

If we substitute the mentioned into equation (6.23) for y we get:

$$y = \frac{b}{\delta} \left[e^{-\beta\tau} \sum_{r=-\infty}^{\infty} c_{2r} \cdot e^{-i2r\tau} \int_0^{\tau} \left(e^{\beta\tau} \cdot e^{jp\tau} \sum_{r=-\infty}^{\infty} c_{2r} \cdot e^{i2r\tau} \right) d\tau - e^{\beta\tau} \sum_{r=-\infty}^{\infty} c_{2r} \cdot e^{2ir\tau} \int_0^{\tau} \left(e^{-\beta\tau} \cdot e^{jp\tau} \sum_{r=-\infty}^{\infty} c_{2r} \cdot e^{-2ir\tau} \right) d\tau \right] \quad (6.24)$$

After integration of expression (6.24) we receive general and partial solutions of Methieu inhomogeneous equation:

$$y = A \cdot e^{\beta t} \sum_{r=-\infty}^{\infty} c_{2r} \cdot e^{2ir\tau} + B e^{-\beta t} \sum_{r=-\infty}^{\infty} c_{2r} \cdot e^{-2ir\tau} +$$

$$+ \frac{b}{\delta} \left[e^{-\beta\tau} \sum_{r=-\infty}^{\infty} c_{2r} \cdot e^{-i2r\tau} \sum_{r=-\infty}^{\infty} \frac{c_{2r}}{2ri + pi + \beta} \cdot e^{(2ri+pi+\beta)\tau} - e^{\beta\tau} \sum_{r=-\infty}^{\infty} c_{2r} \cdot e^{2ir\tau} \sum_{r=-\infty}^{\infty} \frac{c_{2r}}{-2ri + pi - \beta} \cdot e^{(-2ri+pi-\beta)\tau} \right] \quad (6.25)$$

The two first members of the right part of expression (6.25) have arbitrary constants A and B which represent general solution of homogeneous Mathieu equation.

Analyze partial solution of expression (6.25):

$$y = \frac{b}{\delta} \left[\sum_{r=-\infty}^{\infty} c_{2r} \cdot e^{-i2r\tau} \sum_{r=-\infty}^{\infty} \frac{c_{2r}}{2ri + pi + \beta} \cdot e^{(2ri+pi+\beta)\tau} - \sum_{r=-\infty}^{\infty} c_{2r} \cdot e^{2ir\tau} \sum_{r=-\infty}^{\infty} \frac{c_{2r}}{-2ri + pi - \beta} \cdot e^{(-2ri+pi-\beta)\tau} \right]$$

(6.26).

From this expression it is seen that partial solution of Mathieu inhomogeneous equation is limited when $\tau \to \infty$, if:

$$(2ri + pi \pm \beta) \neq 0$$

where r=...-2,-1,0,1,2,3,... is arbitrary integer.

If $(2ri + pi \pm \beta) \to 0$ tend to zero, then fraction $\frac{C_{2r}}{(2ri + pi \pm \beta)} \to \infty$.

Hence follows the resonance condition in the system and it is:

$$2ri + pi \pm \beta = 0 \tag{6.27}$$

Here constant coefficients C_{2r} and β are calculated from Mathieu equations theory or

$$2\pi r + \pi p \mp \bar{D} = 0 \tag{6.28}$$

Here $\bar{D} = \arccos\left[\cos\left(\pi\sqrt{a}\right) + \frac{\pi q^2 \sin\left(\pi\sqrt{a}\right)}{4\sqrt{a}(a-1)}\right]$

$$a = \left(\frac{\omega}{\Omega}\right)^2; \quad q = \frac{k}{2}\left(\frac{\omega}{\Omega}\right)^2; \quad P = \frac{\Theta}{\Omega} = \frac{V_{Sh}}{d\Omega}; \quad r = ...-2,-1,0,1,2,3,...$$

where Ω is guy vibration frequency;

ω pylon vibration frequency;

Θ is wind pulsing frequency.

From equation of resonance (6.28)

$$\frac{V_{Sh}}{d\Omega} = \pm\frac{\bar{D}}{\pi} - 2r$$

Hence, critical velocity of wind when the loss of pylon aerodynamic stability is expected, is equal to:

$$V_{cr.\min} = \frac{d\Omega}{Sh}\left(\pm\frac{\bar{D}}{\pi} - 2r\right) \tag{6.29}$$

where d is pylon diameter:

Sh is Strauhall number for pipe which equals 0.22.

An Example of Practical Realization

Let's consider one support middle pylon of pedestrian guy bridge with dimensions: guy length L=25.5 m; pylon height H=18.0 m; pylon pipe diameter d= 325 mm; guy diameter – 30 mm; pylon bending stiffness $\overline{\overline{D}} = EI = 10586 \cdot 2.1 \cdot 10^6 kg / cm^2$; pipe weight g=66.35 kg/m; cross-section area A=84.52 cm; inertia moment I=10583 cm^4; resisting moment - W=651 sm^3; cross-section radius a pylon $\overline{r} = 16.25 sm$; pipe wall thickness - h=0,85 sm; guy load on pylon origin - $P(t)=P_1-P_0\cos\Omega t=52149,2-4089,99\cos 2\cdot 7,35$; guy vibration frequency - Ω=7,35 1/sec; wind velocity - V=60 m/sec; pylon pipe parameters - $\overline{a}^2 = \frac{h^2}{12\overline{r}^2} = \frac{0,85^2}{12 \cdot 16,25^2} = 0.000228.$

Guy vibration frequency is calculated by formula: pylon guy - Ф 30 mm; guy cross-section area A=4.2151 sm^2 ; guy weight - g=3.85 kg/m; design breaking strength of a guy R=63200 kg (GOST 7669-69*). Guy frequency:

$$\Omega = \frac{1 \cdot 3,14}{25,5}\sqrt{\frac{56239,195}{0,4 \cdot}} = 46,2 \text{ rad/sec } \Omega=7.35\text{Hz}.$$

Parameter $a = \left(\frac{\omega}{\Omega}\right)^2 = \left(\frac{1,923}{7,35}\right)^2 = 0,262^2 = 0,068$.

Parameter $q = \frac{k}{2}\left(\frac{\omega}{\Omega}\right)^2 = \frac{12,75}{2} \cdot 0,068 = 0,4335$.

Pylon vibration frequency:

$$\omega^2 = \frac{\frac{16}{5} \cdot \frac{1}{l^3}}{\frac{r(1-\gamma^2)}{Ea^{-2}}\rho \cdot \frac{104l}{405}} + \frac{p_0 \frac{\overline{r}^2}{\overline{\overline{D}}} \cdot \frac{4}{21} l}{\frac{r(1-\gamma^2)}{Ea^{-2}}\rho \cdot \frac{104l}{405}} =$$

$$= \frac{\frac{16}{5 \cdot 1800^3}}{\frac{16,25(1-0,3^2) \cdot 8 \cdot 10^{-6} \cdot 104 \cdot 1800}{2,1 \cdot 10^{-6} \cdot 228 \cdot 10^{-6} \cdot 405}} + \frac{4089,99\frac{16,25^2 \cdot 4 \cdot 1800}{2,1 \cdot 10585 \cdot 21}}{\frac{16,25(1-0,3^2) \cdot 8 \cdot 10^{-6} \cdot 104 \cdot 1800}{2,1 \cdot 10^{-6} \cdot 228 \cdot 10^{-6} \cdot 405}} =$$

$$= 0,0000039 + 145,89238 = 145,89238$$

$\omega = 12,07859 red / \sec$; ω=1,923 Hz

Parameter R

$$R=\frac{\dfrac{p_0\dfrac{\overline{r}^2}{\overline{\overline{D}}}\cdot\dfrac{4}{21}l}{\dfrac{\overline{r}\left(1-\gamma^2\right)}{Ea^{-2}}\rho\cdot\dfrac{104l}{405}}}{\dfrac{\dfrac{16}{5}\cdot\dfrac{1}{l^3}}{\dfrac{r\left(1-\gamma^2\right)}{Ea^{-2}}\rho\cdot\dfrac{104l}{405}}+\dfrac{p_0\dfrac{\overline{r}^2}{\overline{\overline{D}}}\cdot\dfrac{4}{21}l}{\dfrac{r\left(1-\gamma^2\right)}{Ea^{-2}}\rho\cdot\dfrac{104l}{405}}}=$$

$$=\frac{\dfrac{521149,2\dfrac{16,25^2 4\cdot 1800}{2,1\cdot 10^6\cdot 10583\cdot 21}}{\dfrac{16,25\left(1-0,3^2\right)\cdot 8\cdot 10^{-6}\cdot 104\cdot 1800}{2,1\cdot 10^6\cdot 228\cdot 10^{-6}\cdot 405}}}{\dfrac{\dfrac{16}{5\cdot 1800^3}}{\dfrac{16,25\left(1-0,3^2\right)\cdot 8\cdot 10^{-6}\cdot 104\cdot 1800}{2,1\cdot 10^{-6}\cdot 228\cdot 10^{-6}\cdot 405}}+\dfrac{4089,99\dfrac{16,25^2\cdot 4\cdot 1800}{2,1\cdot 10^6\cdot 10583\cdot 21}}{\dfrac{16,25\left(1-0,3^2\right)\cdot 8\cdot 10^{-6}\cdot 104\cdot 1800}{2,1\cdot 10^{-6}\cdot 228\cdot 10^{-6}\cdot 405}}}=$$

$$=\frac{1860,1929}{145,88796}=12,75$$

Parameter $\overline{D}$

$$\overline{D}=\arccos\left[\cos\left(\pi\sqrt{a}\right)+\frac{\pi q^2\sin\left(\pi\sqrt{a}\right)}{4\sqrt{a}\left(a-1\right)}\right];$$

$$\cos\pi\sqrt{a}=\cos\pi\sqrt{0,068}=\cos 0,26\pi=0,69$$

$$\sin\pi\sqrt{a}=\sin\pi\sqrt{0,068}=\sin 0,26\pi=0,72$$

$$\frac{3,14\cdot q^2}{4\sqrt{a}(a-1)}=\frac{3,14\cdot 0,4335^2}{4\sqrt{0,068}(0,068-1)}=-\frac{0,59}{1,04\cdot 0,932}=-0,6087$$

$$\overline{D}=\arccos\left[0,69-0,4383\right]=\arccos(0,25)$$

$$\overline{D}=0,42\pi$$

Critical velocities of wind (positive $\overline{D}$ case):

$$\mathrm{V}_{cri} = \frac{d\Omega}{Sh}\left(\pm\frac{\overline{D}}{\pi} - 2r\right); r = \ldots -3, -2, -1, 0, 1, 2, 3, \ldots$$

$$\mathrm{V}_{cr1} = \frac{0,325 \cdot 7,35}{0,22}\left(0,42\frac{\pi}{\pi} - 2,0\right) = 4,56m / ces$$

$$\mathrm{V}_{cr2} = 10,85\left(0,42 - 2\left(-1\right)\right) = 26,257m / ces$$

$$\mathrm{V}_{cr3} = 10,85\left(0,42 - 2\left(-2\right)\right) = 47,957m / ces$$

$$\mathrm{V}_{cr4} = 10,85\left(0,42 - 2\left(-3\right)\right) = 69,657m / ces$$

As a result of calculations several critical velocities of wind were detected, among them three velocities of wind are within the wind velocity in bridge territory and only the fourth, $V_{cr4} = 69,657m / sec$ overpasses this limit which denotes that guy bridge pylon and guys need the change of parameters.

Conclusion: the aerodynamic stability of guy bridge pylon is provided if wind critical velocity acting on pylon exceeds 1.5 times the wind velocity on pylon territory.

The comparison of solution of practical examples with the solution of similar tasks for constant parameter systems it follows that not one resonance and, respectively, one critical velocity of wing but infinitely many resonances take place, just some of them having the practical importance (according to amplitude magnitude).

In the systems with periodically changing parameters besides main resonances co-resonances are also detected the vibration amplitudes of which may have particular importance.

Aerodynamic Stability of Prestressed Guy Bridges

The construction of new prestressed guy bridge was developed by Prof. G.Kiziria and eng. B.Maisuradze and patented (P.2620). The mentioned bridge from the right support has prestressed tie-bars which fix stiffening girder. Such bridge is constructed by eng. B.Maisuradze on the river Mtkvari near town Borjomi (see Figure 6.4).

In the work the formulas the frequency of guy bridge bending and twisting vibrations are received. With their use the flutter critical velocity is established when bridge undergoes aerodynamic instability.

1. Determination of Bending Vibrations Frequency of Prestressed Guy Bridge

If we take into consideration the rule of signs then positive bending moment M induces the contraction of upper fibers and stretching of lower fibers. Also it is noticed that negative curvature corresponds to positive bending moment and positive curvature corresponds to negative bending moment.

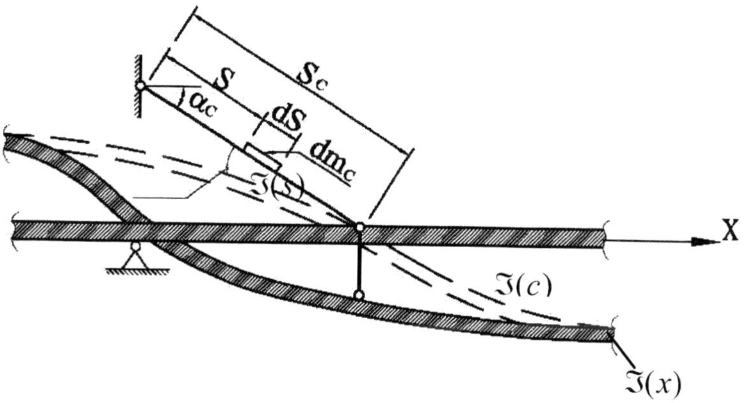

Figure 6.4.

If we denote curvature with x then curvature is reciprocal value of curvature radius.

$$x = \frac{1}{\rho} = \frac{d\theta}{dx} = \frac{d^2y}{dx^2}; \quad \theta \approx \mathrm{tg}\theta \approx \frac{dy}{dx}$$

or $x = \frac{1}{\rho} = -\frac{M}{EI}; \quad \frac{d^2y(x)}{dx^2} = -\frac{M}{EI}$

The exact expression will be:

$$\frac{y''(x)}{\left[1 + y'(x^2)\right]^{3/2}} = -\frac{M}{EI}$$

Use approximate expression:

$$EI\frac{d^2y(x)}{dx^2} = -M \tag{6.30}$$

Use energy conservation rule; the sum of kinetic and potential energies is always constant, i.e.:

W+U=E=const (6.31)

Here W is kinetic energy;
U is potential energy;
E is joint energy.

For free continuous vibrations maximum kinetic energy in tie is observed when potential energy U is equal to zero and vice versa.

According to Rayleigh law maximum values for both energies are equal and are:

$$W_{max}+U_{max}=E \tag{6.32}$$

The characteristic function for vibrating system has the following form:

$$\Phi=U_{max}-\omega^2 W_{max} \tag{6.33}$$

which takes minimum value:

$$d\phi = d(U_{max} - \omega^2 W_{max}) = 0 \tag{6.34}$$

where w is free vibration circular frequency.

Kinetic energy of guy elementary mass is:

$$dW = dm\frac{v^2}{2} \tag{6.35}$$

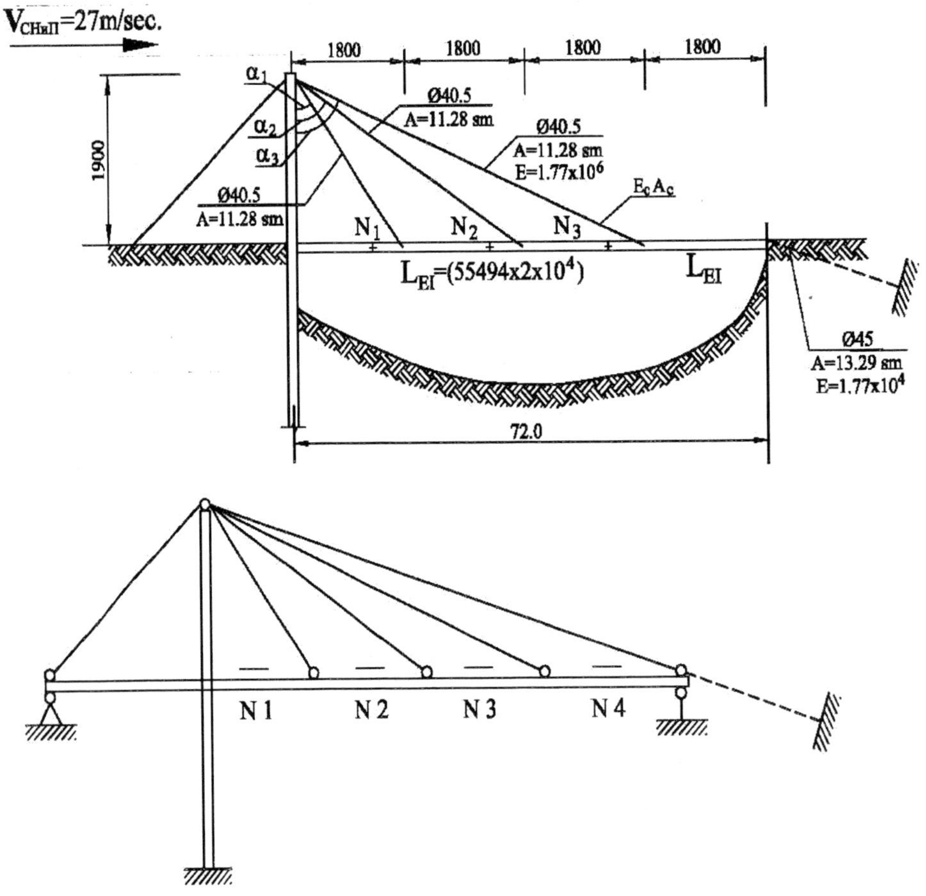

Figure 6.5. Calculation of schemes of cable stayed bridge.

where dm is elementary mass of guy;
v is guy element velocity.
If we represent vibration process as:

$$I = I(x)\sin\omega_v T \tag{6.36}$$

where I(x) is maximum amplitude and T is vibration period then velocity is defined with formula:

$$v = \frac{dI(x,\psi)}{dt} = \omega_v I(x)\cos\omega_v T \tag{6.37}$$

which has maximum value.

When

$$\cos\omega_v T = 1.$$

then

$$v = \omega_v \cdot I(x) \tag{6.38}$$

If we insert (6.38) into expression (6.35) then elementary kinetic energy of a guy will be:

$$dW = \frac{dm}{2}\omega_v^2 I^2(x) \tag{6.39}$$

If stiffness girder is rigidly connected to pylon then kinetic energy can be represented as consisting of two parts:

$$dW = dW_1 + dW_2 \tag{6.40}$$

dW_1 is kinetic energy of stiffness girder:
dW_2 is kinetic energy of guy:

$$dW_1 = \frac{dm}{2}\omega_v^2 I^2(x) = \frac{P}{2g}\omega_v^2 I^2(x)dx \tag{6.41}$$

where P is linear length weight of stiffness girder:

$$dW_2 = \frac{p_c}{v} I^2(s)ds \tag{6.42}$$

here P is linear length weight of guy:

I(s) is vertical displacement of guy element for dS length:

d is free fall acceleration.

From figure 6.5 it is seen that:

$$I(s) = I_c\left(\frac{s}{S_c}\right) \tag{6.43}$$

and if we substitute (6.43) into (6.42) we shall get:

$$dW_2 = \frac{p_c}{2g}\omega_v^2 \frac{I_s^2}{S_c^2} S^2 \tag{6.44}$$

where S_c is the length of inclined guy;

I_s is maximum displacement of stiffness girder fixed to the guy.

Total kinetic energy for guy bridge is determined by integration on the whole length of the bridge:

$$W = \frac{\omega_v^2}{2}\left[\int_L \frac{p}{g} I^2(x)dx + \sum_{c=1}^{m} \frac{p_c}{g}\frac{I_c^2}{S_c^2}\int_0^{S_c} S^2 ds\right] = \frac{\omega_v^2}{2}\left[\int_L \frac{p}{g} I^2(x)dx + \frac{1}{3}\sum_{c=1}^{m}\frac{p_c}{g} I_c^2 S_c\right] = \omega_v^2 \overline{W} \tag{6.45}$$

Potential energy of guy bridges may be calculated by the following formula:

$$U = \frac{1}{2}\int_L PI(x)dx \tag{6.46}$$

or

$$U = V_1 + V_2 + V_3 \tag{6.47}$$

The expression of bending moment in guy bridge stiffness beam with consideration of linear forces is:

$$M = -EI\frac{d^2 I(x)}{dx^2} + \sum_{i=1}^{k} N_i I(x) - \sum SI(x) \tag{6.48}$$

where $\sum_{i=1}^{k} N_i$ is the sum of compressive forces;

S is tensile force induced by tie-bar in stiffness beam;

V_1 is effort energy in stiffness beam;

V_2 is effort energy in guys;
V_3 is internal work in guys executed with guy forces on second order deformations.
Potential energy of stiffness beam is:

$$+V_1 = \frac{1}{2}\int_L \frac{M^2}{EI}dx = \frac{1}{2EI}\int_L \left[-EI\frac{d^2 I(x)}{dx^2} + \sum_{i=1}^{k} N_i I(x) - \sum SI(x) \right]^2 dx \tag{6.49}$$

where EI is girder bending stiffness.

In case of vibration forces X_c increase and decrease in guys and the internal work is calculated with formula:

$$V_2 = \frac{1}{2}\sum_{c=1}^{m}\frac{X_c^2 S_c}{E_c A_c} \tag{6.50}$$

E_c is guy elasticity module;
A is cross-section area of guy.
Guy stretching which causes maximum displacement is equal to:

$$\Delta S_c = I_c \sin\alpha_c \tag{6.51}$$

on the other hand:

$$\Delta S_c = \frac{X_c S_c}{E_c A_c} \tag{6.52}$$

The force in guy is determined:

$$X_c = I_c \frac{E_c A_c}{S_c}\sin\alpha_c \tag{6.53}$$

where α_c is guy incline angle to horizon.

Finally we get

$$V_2 = \frac{1}{2}\sum_{c=1}^{m}\frac{E_c A_c I_c^2 \sin\alpha_c}{S_c} \tag{6.54}$$

At vibrations force changes in a guy can be represented as:

$$X_{cv} = X_c \sin(\alpha_c + d\alpha_c) - X_c \sin\alpha_c$$

As we can assume that $\cos d\alpha_c \approx 1$ and $\sin d\alpha_c \approx d\alpha$ then we shall receive:

$$X_{cv} = X_c \cos\alpha_c \cdot d\alpha \tag{6.55}$$

Angle of rotation of a guy is:

$$d\alpha = \frac{I \cos\alpha_c}{S_c} \tag{6.56}$$

If we insert (6.56) into expression (6.55) we get:

$$X_{cv} = X_c \frac{\cos^2\alpha}{S_c} I \tag{6.57}$$

Work which is executed with internal forces of a guy on the second order deformations is expressed as:

$$V_3 = \sum_{c=1}^{m} \int_0^{l_c} X_{cv} \cdot dI = \sum_{c=1}^{m} \int_0^{l_c} \frac{X_{cv} \cos^2\alpha_c \cdot I \cdot dI}{S_c} \tag{6.58}$$

or after integrating of (6.58) we get:

$$V_3 = \frac{1}{2} \sum_{c=1}^{m} X_c \frac{\cos^2\alpha_c}{S_c} I_c^2 \tag{6.59}$$

or if we insert X expression into (6.59) we shall get:

$$V_3 = \frac{1}{2} \sum_{c=1}^{m} \frac{E_c A_c \sin\alpha_c \cos^2\alpha_c I_c^3}{S_c^2} \tag{6.60}$$

Now, if we sum up kinetic and potential energies and equalize them to each other, we get:

$$dW_1 + dW_2 = V_1 + V_2 + V_3 \tag{6.61}$$

Hereof free vibrations frequency is equal to:

$$\omega_c^2 = \frac{U}{W} = \frac{V_1 + V_2 + V_3}{dW_1 + dW_2} \tag{6.62}$$

or after substitution we get:

$$\frac{\omega_v^2}{2}\left[\int_L \frac{p}{g} I^2(x)dx + \frac{1}{3}\sum_{c=1}^{m}\frac{p_c}{g} I_c^2 S_c\right] = \frac{1}{2EI}\left\{\int_0^L \left[-EI\frac{d^2 I(x)}{dx^2} - 2SI(x) + \sum_{i=1}^{k} N_i I(x)\right]^2\right\} dx + \tag{6.63}$$
$$+\frac{1}{2}\sum_{c=1}^{m}\frac{E_c A_c I_c^2 \sin\alpha_c}{S_c} + \frac{1}{2}\sum_{c=1}^{m}\frac{E_c A_c \sin\alpha_c \cos^2\alpha_c I_c^3}{S_c^2}$$

If we assume the vibration mode as:

$$I(x) = I_c \sin\frac{n\pi x}{2} \tag{6.64}$$

and substitute into expression (6.63) finally we shall have:

$$\frac{\omega_v^2}{2}\left[-\frac{P}{g}\cdot\frac{L}{2} + \frac{1}{3}\sum_{c=1}^{m}\frac{P_c}{g}S_c\right] = -\frac{EIn^4\pi^4}{4L^3} - \frac{n^2\pi^2}{4L}2S + \frac{n^2\pi^2}{2L}\sum_{i=1}^{k}N_i + \frac{4S^2L}{2EI} - \frac{2S\sum_{i=1}^{k}N_i L}{2EI} + \frac{\sum_{i=1}^{k}N_i^2 L}{2EI} \tag{6.65}$$
$$+\frac{1}{2}\sum_{c=1}^{m}\frac{E_c A_c \sin\alpha_c}{S_c} + \frac{1}{2}\sum_{c=1}^{m}\int_0^{l_c}\frac{E_c A_c \sin\alpha_c \cos^2\alpha_c I_c^3}{S_c^2}$$

After canceling and simplifying we get:

$$-\frac{\omega_v^2}{2}\left[\frac{P}{g}\cdot\frac{L}{2} - \frac{1}{3}\sum_{c=1}^{m}\frac{P_c}{g}S_c\right] = -\left(\begin{array}{l}\frac{n^4\pi^4 EI}{4L^3} + \frac{n^2\pi^2 S}{2L} - \frac{n^2\pi^2}{2L}\sum_{i=1}^{k}N_i - \frac{4S^2L}{2EI} + \frac{2S\sum_{i=1}^{k}N_i L}{EI} - \frac{\sum_{i=1}^{k}N_i^2 L}{2EI} - \\ -\frac{1}{2}\sum_{c=1}^{m}\frac{E_c A_c \sin\alpha_c}{S_c} - \frac{1}{2}\sum_{c=1}^{m}\int_0^{l_c}\frac{E_c A_c \sin\alpha_c \cos^2\alpha_c}{S_c^2} I_c^3\end{array}\right)$$

If we neglect second order deformations in a guy the last member will drop out:

$$\frac{\omega_v^2}{2} = \frac{\frac{n^4\pi^4 EI}{2L^3} + \frac{n^2\pi^2 S}{L} - \frac{n^2\pi^2}{L}\sum_{i=1}^{k}N_i - \frac{4S^2L}{2EI} + \frac{4S\sum_{i=1}^{k}N_i L}{EI} - \frac{\sum_{i=1}^{k}N_i^2 L}{2EI} - \sum_{c=1}^{m}\frac{E_c A_c \sin\alpha_c}{S_c}}{\frac{PL}{2g} - \frac{1}{3}\sum_{c=1}^{m}\frac{P_c}{g}S_c}$$

Dimensional analysis:

$$\frac{1}{sec^2} = \frac{\frac{\frac{kg}{m^2}m^4}{m^3} + \frac{kg}{m} - \frac{kg}{m} - \frac{kg^2 m}{\frac{kg}{m^2}m^4} + \frac{kgkgm}{\frac{kg}{m^2}m^4} - \frac{kg^2 m}{\frac{kg}{m^2}m^4} - \frac{\frac{kg}{m^2}m^2}{m}}{\frac{\frac{kg}{m}}{sec^2} - \frac{\frac{kg}{m}}{sec^3}} = \frac{\frac{kg}{m}}{\frac{kg}{m}sec^2}$$

Calculation formula:

$$\frac{\omega_v^2}{2}=\frac{\dfrac{n^4\pi^4 EI}{2L^3}+\dfrac{2n^2\pi^2 S}{L}-\dfrac{n^2\pi^2}{L}\sum_{i=1}^{k}N_i-\dfrac{4S^2L}{2EI}+\dfrac{2S\sum_{i=1}^{k}N_iL}{EI}-\dfrac{\sum_{i=1}^{k}N_i^2L}{2EI}-\sum_{c=1}^{m}\dfrac{E_cA_c\sin\alpha_c}{S_c}}{\dfrac{PL}{2g}-\dfrac{1}{3}\sum_{c=1}^{m}\dfrac{P_c}{g}S_c}$$

P=q=kg/m=kN/m – a beam;
P_c=q_c=kg/m=kN/m – a beam;
L is beam span, m;
S_c is guy length, m;
A_c is guy cross-section area, cm^2

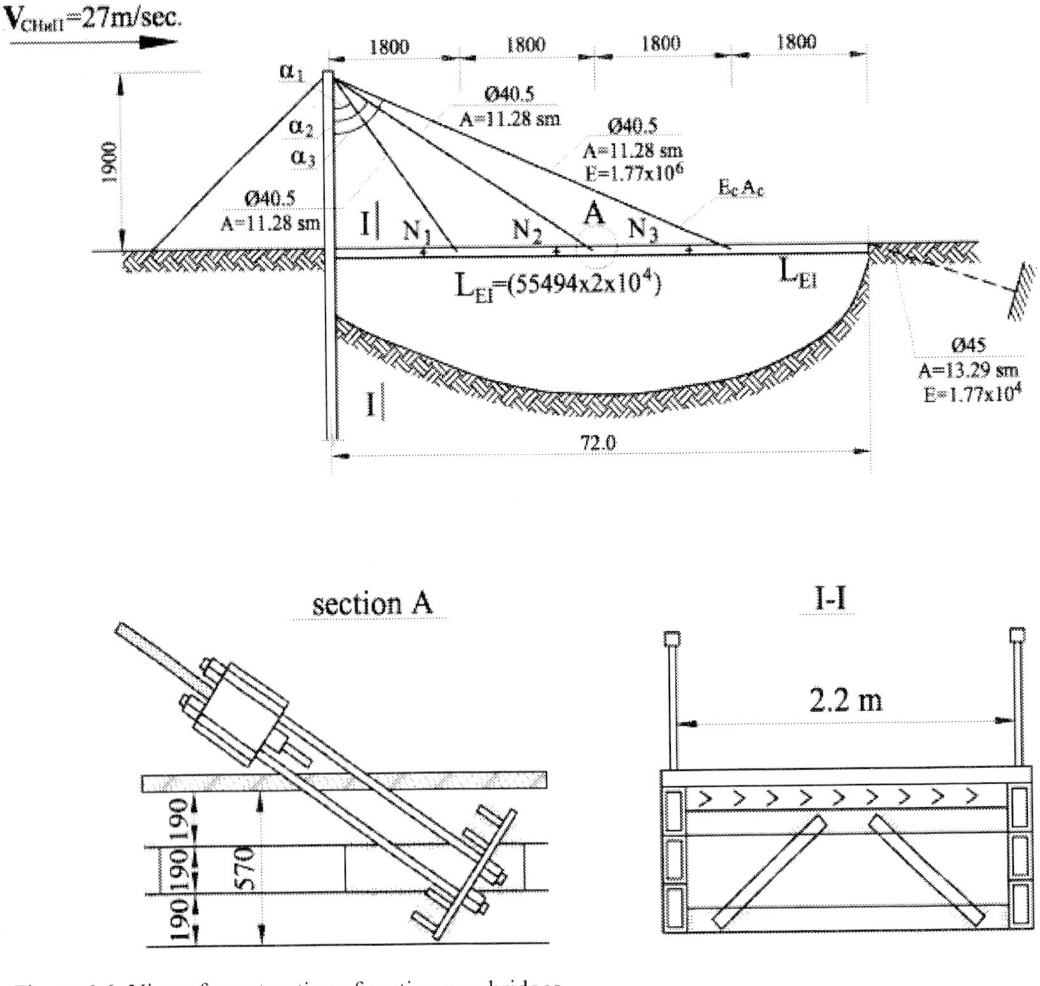

Figure 6.6. View of construction of section guy bridges.

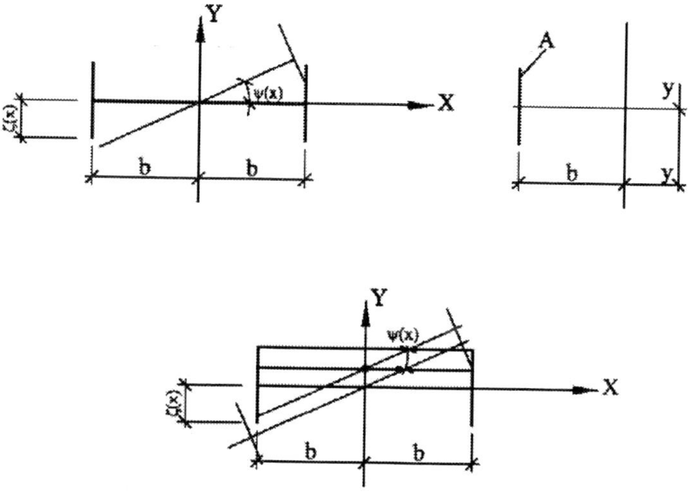

Figure 6.7. Deformation of twisting vibrations frequency of guy bridges.

If we denote guy bridge cross-section rotation angle with $\Psi(x)$ and connect bending and twisting deformations then:

$$m(x) = b\psi(x) \tag{6.66}$$

where 2b is stiffness beam width (Figure 6.7)

The mentioned ψ(x) is periodical maximum of an angle when bridge cross-section rotates by abscissa x.

If we take relation $\frac{I(x)}{b} = tg\psi(x) = \psi(x)$, we get the above given expression $I(x) = b\psi(x)$. Here we also give inertia moment expression at twisting from Figure - $I\ = A\ \cdot b^2$; the stiffness beam which undergoes free vibrations (San-Venan torsion), "rotation" or angle turn for unit length of bridge is:

$$\frac{d\psi(x)}{dx} = \frac{M_g}{gI_g} \tag{6.67}$$

If we insert relation (6.66) into the expressions of bending vibrations kinetic and potential energies, we get the expression of torsion vibrations kinetic energy with consideration of prestressing in stiffness beam.

Expression of kinetic energy is:

$$W_g = \frac{\omega_g^2}{2}\left\{\frac{P}{g}\int_0^L I^2(x)dx + \sum_{c=1}^{m}\frac{p_c}{g}\frac{I_c^2}{S_c^2}\int_0^{S_c} S^2 ds\right\} = \frac{\omega_g^2}{2}\left[\frac{P}{g}\int_0^L I^2(x)dx + \frac{1}{3}\sum_{c=1}^{m}\frac{p_c}{g}I_c^2 S_c\right] =$$

$$= \frac{\omega_g^2}{2}\left[\frac{P}{g}\int_0^L b^2\psi^2(x)dx + \frac{1}{3}\sum_{c=1}^{m}\frac{p_c}{g}b^2\psi_c^2(x)S_c\right] = \tag{6.68}$$

$$= \frac{\omega_g^2}{2}\left[\frac{P}{g}\int_0^L \frac{I_g}{A_g}\psi^2(x)dx + \frac{b}{3}\sum_{c=1}^{m}\frac{p_c}{g}\psi_c^2(x)S_c\right] = \omega_g^2\overline{W}_g$$

Potential energy expression at torsion vibrations with consideration of prestressing in stiffness beam is:

$$V_g = V_1 + V_2$$

Potential energy expression at torsion is:

$$V_1 = \int_0^L \frac{M_g^2 dx}{2GI_g}, \quad M = GI\,\frac{d\psi(x)}{dx} - \frac{1}{2}b\left(\sum N + \sum S\right)\psi(x) \tag{6.69}$$

$$\psi(x) = \psi_c \sin\frac{n\pi x}{L} \tag{6.70}$$

or

$$M = \left[GI\,\frac{d\psi(x)}{dx} - \frac{1}{2}b\left(\sum N + \sum S\right)\psi(x)\right]^2$$

Here of:

$$V_1 = \frac{1}{2GI_g}\int_0^L\left[GI_g\frac{d\psi(x)}{dx} - \frac{1}{2}b\left(\sum N + \sum S\right)\psi(x)\right]^2 dx =$$

$$= \frac{1}{2GI_g}\int_0^l\left[\left(GI_g\right)^2\left(\frac{d\psi(x)}{dx}\right)^2 - 2GI_g\frac{d\psi(x)}{dx}\cdot\frac{1}{2}b\left(\sum N + \sum S\right)\psi(x) + \frac{1}{4}b^2\left(\sum N + \sum S\right)^2\psi^2(x)\right]dx =$$

$$= \frac{1}{2GI_g}\int_0^l\left(GI_g\right)^2\cdot\frac{n^2\pi^2}{L^2}\cdot\cos^2\frac{n\pi x}{L}\cdot\psi_c^2 dx -$$

$$-\frac{1}{2GI_g}\cdot 2GI_g\int_0^l\frac{d\psi(x)}{dx}\psi(x)\frac{1}{2}b\left(\sum N + \sum S\right)dx + \frac{1}{2GI_g}\frac{1}{4}b^2\left(\sum N + \sum S\right)^2\psi^2(x)\cdot\frac{L}{2}\psi_c^2 =$$

$$= \frac{GI_g n^2\pi^2}{4L}\psi_c^2 + \frac{b^2\left(\sum N + \sum S\right)^2}{16GI_g}\psi_c^2$$

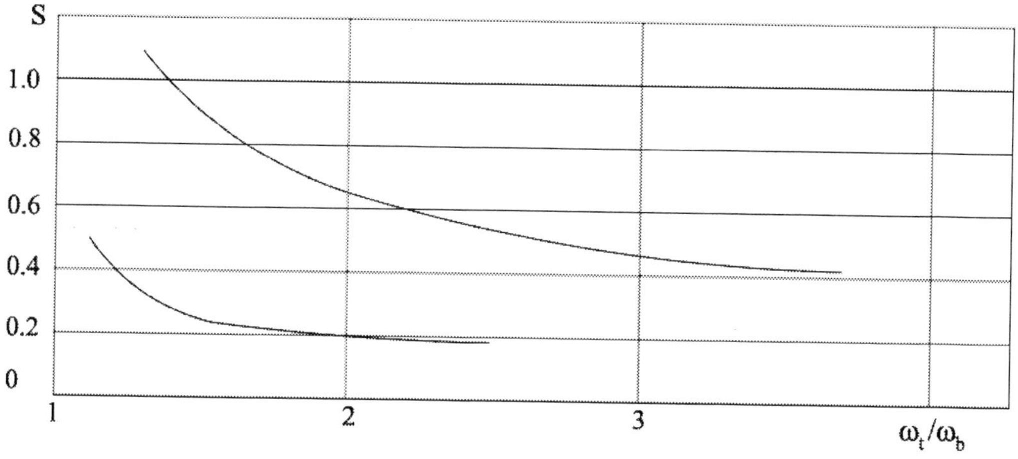

Figure 6.8. Parameter S is determined according $\omega_{tw} / \omega_{be}$

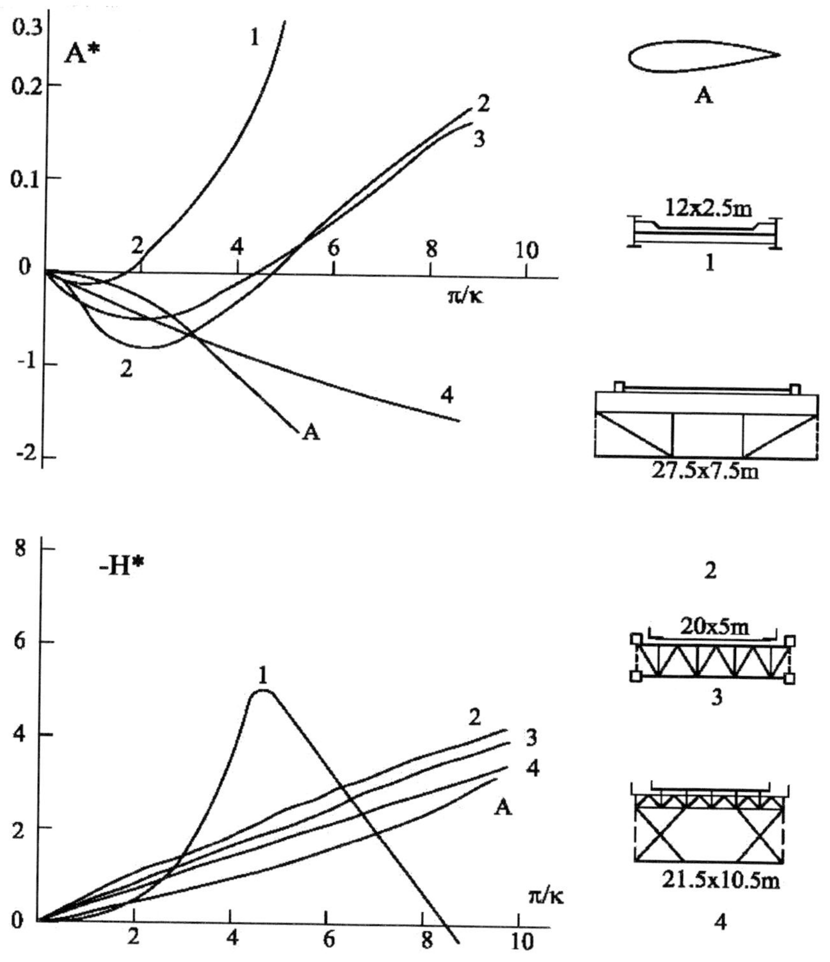

Figure 6.9. A_i^* and H_i^* coefficients of different cross-sections of a guy bridge.

Thus:

$$V_1 = \frac{GI_g n^2 \pi^2}{4L} \psi_c^2 + \frac{b^2 \left(\sum N + \sum S\right)^2 \cdot \psi_c^2}{16 GI_g}$$

$$V_2 = \frac{1}{2} \sum \frac{E_c A_c I_c^2 \sin \alpha_c}{S_c} = \frac{b^2}{2} \sum \frac{E_c A_c \psi_c^2 \sin^2 \alpha_c}{S_c}$$

Where: ω_{tw} -frequency of twisting vibration

ω_{be} -frequency of bending vibration

Totally the expression of potential energy will be:

$$V_g = V_1 + V_2 = \frac{GI_g n^2 \pi^2}{4L} \psi_c^2 + \frac{b^2 \left(\sum N + \sum S\right)^2 \cdot L}{16 GI} \psi_c^2 + \frac{b^2}{2} \sum \frac{E_c A_c \sin^2 \alpha_c}{S_c} \psi_c^2$$

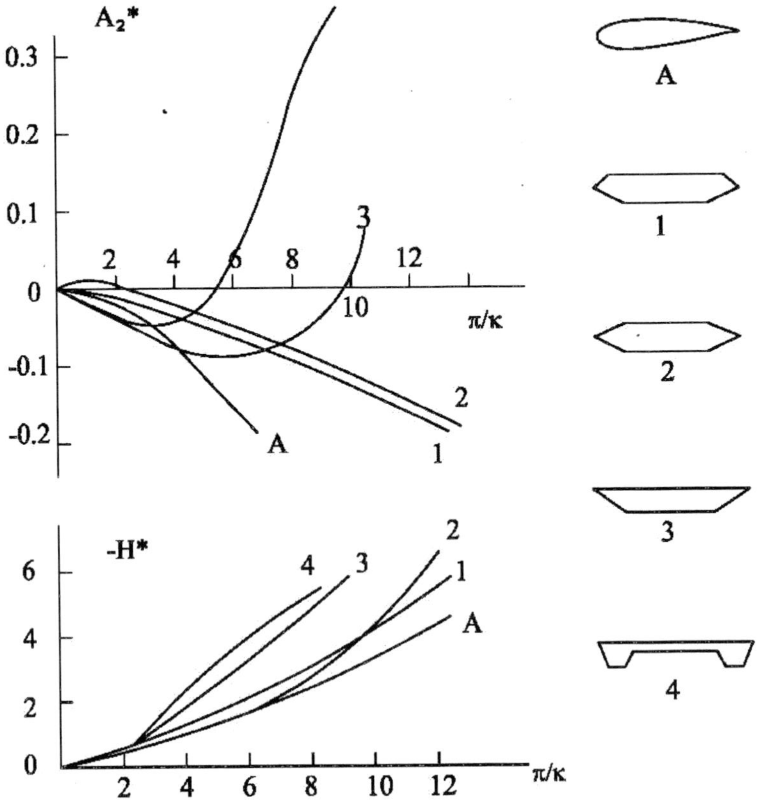

Figure 6.10. A_2^* and H^* coefficients of box like stiffness beams of a guy bridge.

According to energetic law maxW-maxV=0:

$$\frac{\omega_g^2}{2}\left[\frac{P}{g}\int_0^L\frac{I_g}{A_g}\psi^2(x)dx+\frac{b}{3}\sum_{c=1}^{m}\frac{p_c}{g}\psi_c^2(x)S_c\right]=\frac{\omega_g^2}{2}\left[\frac{P}{g}\frac{I_g}{A_g}\frac{L}{2}\psi_c^2+\frac{b^2}{3}\sum\frac{p_c}{g}\psi_c^2S_c\right]$$

Insert into equation:

$$\frac{\omega_g^2}{2}\left[\frac{P}{g}\frac{I_g}{A_g}\frac{L}{2}\psi^2+\frac{b^2}{3}\sum\frac{p_c}{g}\psi_c^2S_c-\right]-\frac{GI_gn^2\pi^2}{4L}\psi_c^2+\frac{b^2\left(\sum N+\sum S\right)^2\cdot L}{16GI_g}\psi_c^2+$$
$$-\frac{b^2}{2}\sum\frac{E_cA_c\sin\alpha_c}{S_c}\psi_c^2=0$$

Canceling by ψ_c^2 we have:

$$\frac{\omega^2}{2}\left[\frac{P}{g}\frac{I_g}{A_g}\frac{L}{2}+\frac{b^2}{3}\sum\frac{p_c}{g}S_c\right]-\frac{GI_gn^2\pi^2}{4L}-\frac{b^2\left(\sum N+\sum S\right)^2\cdot L}{16GI_g}-\frac{b^2}{2}\sum\frac{E_cA_c\sin\alpha_c}{S_c}=0$$

Frequency of torsion vibrations of bridge stiffness beam is:

$$\omega_t^2=\frac{\dfrac{n^2\pi^2GI_g}{4L}+\dfrac{b^2\left(\sum N+\sum S\right)^2\cdot L}{16GI_g}+\dfrac{b^2}{2}\sum\dfrac{E_cA_c\sin\alpha_c}{S_c}=0}{\dfrac{P}{g}\dfrac{I_g}{A_g}\dfrac{L^2}{2}+\dfrac{b^2}{3}\sum\dfrac{p_c}{g}S_c} \qquad (6.71)$$

Dimensional analysis of formula (6.71) is:

$$\frac{1}{sec^2}=\frac{\dfrac{kg\cdot m^2}{m}+\dfrac{m^2\cdot kg^2\cdot m}{kg\cdot m^2}+\dfrac{m^2\cdot kg}{m}}{\dfrac{kg\cdot \sec^2\cdot m^4\cdot m}{m\cdot m\cdot m^2}+\dfrac{m^2\cdot kg\cdot \sec^2\cdot m}{m\cdot m}}=\frac{kg\cdot m}{kg\cdot \sec^2\cdot m}=\frac{1}{\sec^2}$$

Here an assumption is made: because of low incline angle of a guy and a tie-bar the forces in guys and tie-bars are the same as in stiffness beam. The reaction of guys and tie-bars opposite to twisting moment create restoring torsion moment with opposite sign.

Module of torsion elasticity is $G=\dfrac{E}{2(1+\mu)}$; μ - is Poisson coefficient; E is the module of elasticity; ψ_c is the rotation angle of bridge cross-section in the points of guy and tie-bar fixing.

Technical torsion frequency is $n = \frac{\omega_t}{2\pi}$

Testing on Flutter

$$\omega_f^2 = \omega_t^2 \frac{\left(1 - \frac{\omega_b^2}{\omega_t}\right)\frac{4b^2 \cdot A_2^*}{\tau^2 H_1^*}}{1 - \frac{4b^2 \cdot A_2^*}{\tau^2 H_1^*}}$$

Critical frequency of flutter is:

$$V_f = \frac{\omega_f \cdot b \cdot 2}{S}$$

Here τ is inertia radius of stiffness beam cross-section.

A_2^*, H_1^* and S are obtained from graphs according to π /k (see Figure 6.8, 6.9 and 6.10).

The above mentioned formulas are proposed by Prof. A.Petropavlovski and Prof. A.Potapkin, their methods being used for testing aerodynamic stability of bridge stiffness beam.

Experimental determination of guy bridge natural frequencies vibrations?

Testing of pedestrian guy bridge was done in situ in Borjomi (Georgia) according to the above given parameters. The meters installed on the guy bridge was fixing vibrations in four points (N 1, N 2, N 3 and N 4) (Figure 6.5).

On Figure 6.5. the places of installation of meters on the guy bridge.

In the first version four points were the places of fastening of guys, in the second case the points were in the middle of spans between guys.

The disposition of diagrams given on oscillograms correspond to the numbers N 1, N 2, N 3 and N 4 of the location of meters.

Time mark on oscillogram is 0.1 sec.

As the analysis of the diagram shows natural vibrations frequency recorded by the meter located in point N 4 was ω_1=1,36 1/sec; analogously are in points N 2 and N 3. As to point N 1, here the record has two frequencies, one is 4.0 1/sec, that is conditioned by the nearness of pylon fixing point.

In all points displacements are synchronous which proves that here the first type of vibrations have taking place.

In case of meters locating at guy supports the same results are received except for point N 1 where ω_1=2.5 1/sec on which higher frequencies are applied.

The logarithmic coefficient of frequency damping is calculated with the following formula:

$$\delta = \frac{1}{k}\ln\left(\frac{x_1}{x_{1+k}}\right) = \frac{1}{2}\ln\left(\frac{3,5}{1}\right) = \frac{1}{12}\ln 3,5 = 0,1044$$

By According to Prof. Kazakevich this coefficient varies from 0.015□0.0027 to 0.075.

The frequencies of natural vibrations received for guy bridges at bending vibrations are 1.36 1/sec, 2.5 1/sec and 4.0 1/sec which are close to theoretically received result – 3.53 1/sec (Figure 6.11).

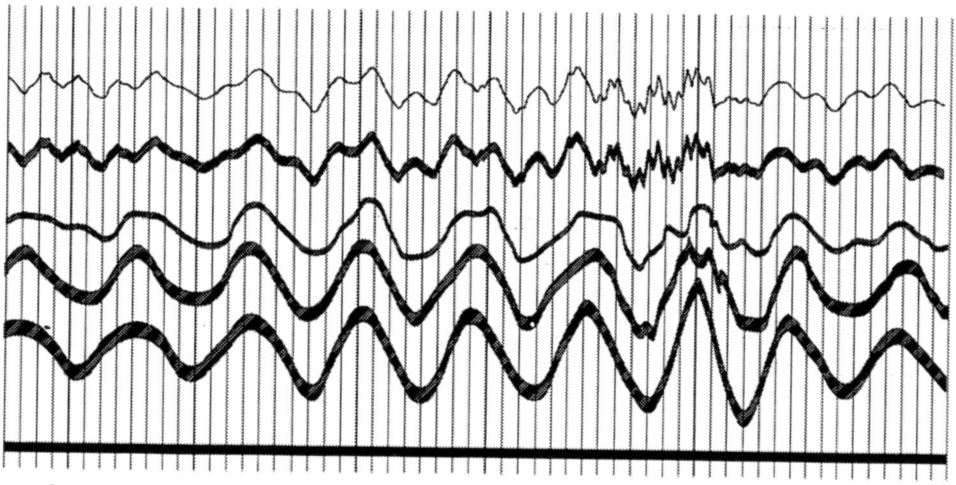

Figure 6.11. Oscillogram of natural vibration of a guy bridge sliding beam in Borjomi (Georgia).

Chapter 7

Application of Electromechanical and Fiber-optical Sensors in the Management of Space Structures Operation

7.1. Cable, Film and Net Space Structures

At present the application of different purpose structures (sounding cable, tow line, conveyer belt, solar belt battery, solar reflector, cosmic antennas, solar sail, etc.) in space conditions are being tested.

These structures are to satisfy such criteria as: easy transportation to earth orbit, small mass, folding and unfolding properties, tight packing. All this is possible only in structures with prestressed elements (cables, nets, films).

In October 1984 in Lausanne (Switzerland) at the XXXV Congress of Astronautics International Federation a special section "Application of cables in space" was founded where the possibilities of using cables in space have been considered.

Practically cables were first used in space in 1965 when astronaut A.Leonov came out into open space and was tied to spaceship "Voskhod-2" with a cable.

In 1967 when American spaceships "Jeminy-11" and "Jeminy-12" where in cosmos the movement of the spaceship was imitated when spaceship was connected to the end stage of the launch vehicle with a cable.

In 1984 a program was set up by NASA and European Cosmic Agency about realization of three flights on the basis of orbital spaceship using tied satellite systems.

The first such flight was planned in 1987-1990 for testing 20 km length electromagnetic cable system.

The second test implies atmospheric sounding with 100 km length cable dropped down from the height of 120 km.

For such cases the allowed length of cable is:

$$L_* = 2\sigma_* / \left(3\rho_M \omega^2\right)$$

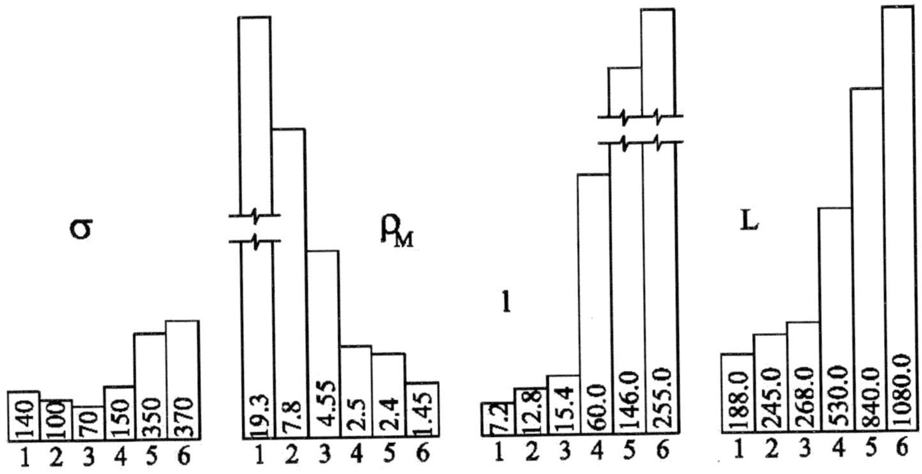

Figure 7.1. Cable characteristics.

where $\sigma_* = T_* / f$ is tension in cable at some support height; ρ_M is cable material density; f is cable cross-section area; T_* is cable strain; ω is orbital angular velocity.

In earth conditions the analog to the above formula is value $l_* = \sigma_* / \rho_M g$ - cable length proceeding from the allowable strength condition. (Figure 7.1)

There are several possible versions of the project of cable connection of spaceships and satellites. For example, 150 kg secondary satellite is coupled to the spacecraft – clum special shuttle - with two cables; stainless steel of 0.5 mm diameter is used for cable material (Figure 7.2).

The orbit height of the main spacecraft is 220 km, the distance between the points of cable securing to spacecraft is 10 m. The cable has two initial lengths: 1 km and 100 km. The stretching force in cables differs respective to initial lengths (stretching force of 1 km long cable is 0.63 N, that of 100 km long cable – 31.56 N). At cable take up when cable length achieves 100 km, single cable stretching force will twice exceed double cable stretching force.

At present in order to receive power energy for cosmic stations solar batteries are widely used. They are of various kinds. Roll type solar film batteries in initial state coiled on a drum are of particular interest. The batteries retain unfolded state by means of tightening mechanism. FRUSA is a roll construction solar battery (Figure 7.3) which was designed for "Agena" rocket (USA) stage and was launched to polar orbit in 1971 (height - 640 km).

This solar battery consists of two panels with dimensions 1.7x4.9 m coiled on 20 cm diameter single common drum.

In 1929 German scientist Herbert Obert expressed the idea about creation such cosmic reflectors which would direct sun rays to the Earth surface. This idea afterward was developed by American scientist Kraft Erik.

The prolongation of day light in big cities for several hours, illumination of streets, transport mains and building sites with solar satellite reflectors is economically justified. It is calculated that power energy economy received at illumination of 5 such cities as Moscow will in 4-5 years repay the expenses made for construction of such reflectors.

The use of solar satellite reflectors is also advisable at harvesting or sowing in night hours, at great earthquakes or other natural disasters when rescue operations are carried out.

The researchers generally conceive the construction of cost efficient orbital satellite-reflectors design as self-sufficient systems consisting of separate groups located at different orbits (Figure 7.4). Each of them represents a usual folded umbrella which after getting to the orbit unfolds automatically.

Reflector surface which reflects solar rays will be polymer aluminized film. Light flux orienting will be done with a special control system.

The Soviet scientists, academicians Zh.Alferov and V.Kantor think that "experimental samples of space reflectors may be constructed in the nearest decade."

The present day cosmic energy is mainly solar energy. It is already over 30 years since solar semiconductor batteries are supplying spacecrafts with power but their capacity is low, just some kilowatts. The construction of large industrial complexes in space needs the increase of power energy consumption and earth supply with power energy at the expense of solar energy. The specialists consider solar cosmic power stations as such systems.

In 1975 "Boying" Corp. has developed the project of space solar thermal power station with aggregate modules.

Bearing structure of solar power station delivered into space in folded state represents truss like elements which unfold on orbit by means of cables. One such solar concentrator contains 52000 film facets. Facet is a hexagonal light frame with aluminum plated film stretched over it. The distance between parallel ribs of bevel is 18 m (Figure 7.5). Each bevel has an independent director in order that sun rays were constantly oriented to solar batteries or to solar power station receiver.

In aggregate complex facet dimensions in plan achieve 2057x2910 m and mass is 42000 tons.

1 kW power energy price received in such a way is 10000 US $.

In constructional view-point space net antennas are close to solar concentrators (Figure 7.6).

In accordance with the present demands space antenna diameter achieves 10-30 m. Large diameter antennas consist of several sections which unfold independently. According to the structure there exist rigid-, folding- and elastic-ribbed umbrella-like antennas. Rigid-ribbed antenna is unfolding by means of electric motor on the drum of which cable is reeled, the latter being connected with all ribs. When cable is reeled on the drum the ribs abut to each other and mirror surface begins to unfold. The springs are used for folding, they separate ribs and the mirror is folded.

Elastic-ribbed antenna is very compact in packed state. Because of low rigidity it is used only in space conditions. Elastic ribs are looped around central rod which unfolds as a telescope. Such antennas are mainly developed by the "Lokheed missiles and space" Corp. (USA). Antenna unfolds due to flexible properties of elastic ribs. It is stated that ribs elasticity is sufficient for unfolding 23 m diameter antenna; for antennas with larger diameter additional mechanisms are needed.

ATS -6 with diameter 9.1 m and 48 elastic ribs belong to such antennas. In folded state container diameter is 2 m, height - 0.2 m. NASA has developed the antenna "riser-collar" with diameter of 64 m; its surface is of membrane type and is to retain practically parabolic form.

Aureate molybdenum knitted net with 0.25 mm cells is used for antenna surface.

The first practical laboratory tests of solar sail technology were started in 1969-1970 in the USA.

Square-in-plan solar sail is called heliorotor. This is enormously large film screen stretched over truss-like frame of 850x850 m. It is made of polyamide aluminum plated plastic material (captone) with thickness not more than 2.5 mkm. The thickness of aluminum plating is 0.1 mkm.

The construction looks like a "kite". The frame by means of the central rod is fixed with braces. The mass of container is approximately 800 kg, that of the whole construction - 5000 kg (Figure 7.7).

Naturally, the discussed projects are the tomorrow of astronautics which is not very far.

Unfolding band constructions and stretch device are given in Figure 7.13 and Figure 7.14.

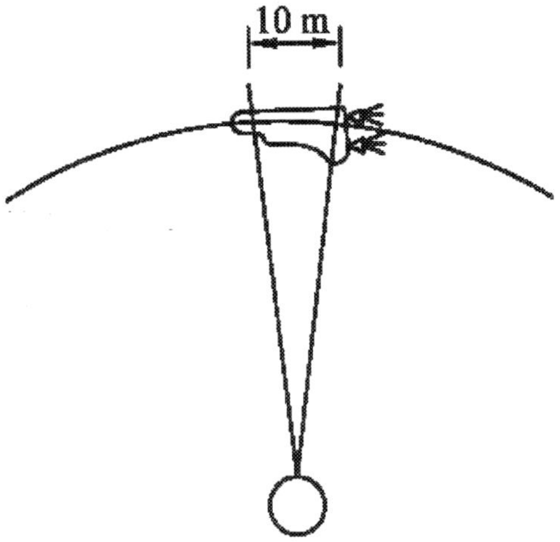

Figure 7.2. Cable tow.

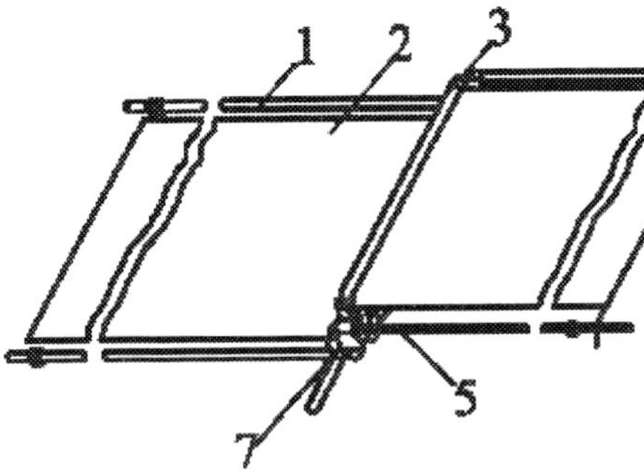

Figure 7.3. Diagram of roll type solar battery: 1 – unfolding pipe-like beam; 2 – panel of solar element; 3 – hoisting roller; 4 – compensator of beam lengthening; 5 – drum; 6 - beam motor; 7 – cantilever; 8 – tension cross-piece.

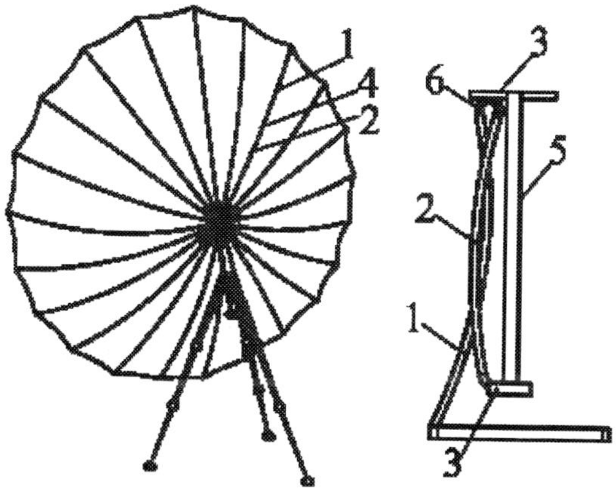

Figure 7.4. Unfolded solar reflector. 1, 2 – bends; 3 – elastic film spring 4 - middle point of ribs; 5 – central stem; 6 – special loop.

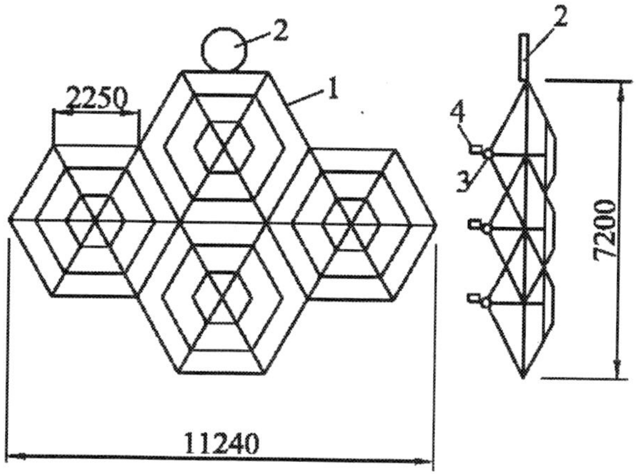

Figure 7.5. Four module space solar power station diagram: 1 – facet concentrator; 2 – antenna; 3 – receiver-transformer; 4 – refrigerating reflector.

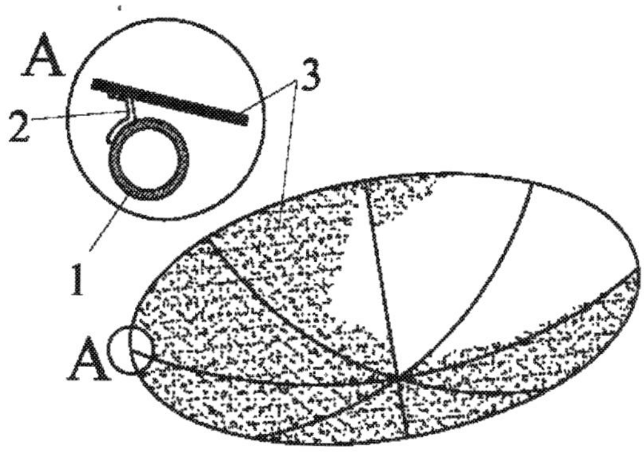

Figure 7.6. Rigid multi-section aluminized concentrator: 1 – toroid; 2 – circular cantilever: 3 – concentrator shell section.

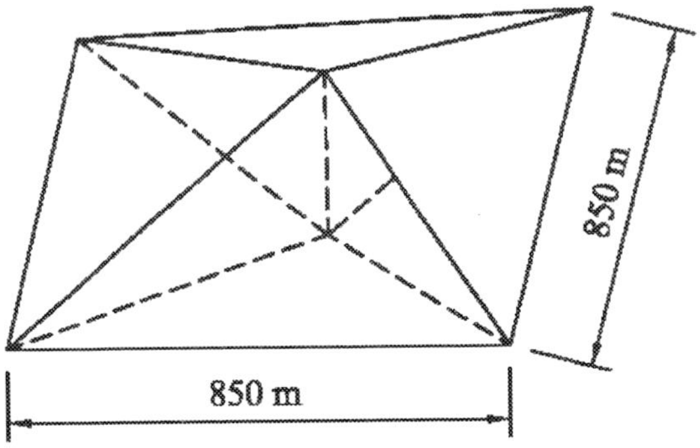

Figure 7.7. Square solar sail.

7.2. Regulation of Natural Vibration Frequencies of Complex Configuration Membranes

In a number of works natural vibrations of rectangular and circular configuration membranes fixed in contour on the whole perimeter are considered.

Often it becomes necessary to define natural vibration frequencies of membranes in complex configuration plan and supported in discrete points of the profile.

The solution of the mentioned problem for membranes with complex outline in plan and supported in discrete points of profile when the rest sections of the membrane are freely hanging is a complex task as boundary conditions become inhomogeneous and the solution of the problem in radicals is practically impossible.

The solution of the problem in opportune and practically acceptable form is known for the circle and thus the complex profile membrane can be conformably restructured for a circle.

In the new area Poisson equation in plane (u, v) will acquire the following form:

$$\nabla^2 W = \frac{1}{\left|W'(\xi)\right|^2} \frac{q(u,v)}{S} \tag{7.1}$$

where $\frac{1}{\left|W'(\xi)\right|^2}$ is distortion factor;

$q(u,v)$ is uniformly distributed load on unit area of membrane;

S is membrane tension;

The boundary conditions will be:

$$W\Big|_{A',B',\ldots} = 0; \qquad W\Big|_{\Gamma_1',\Gamma_2',\ldots} = 0;$$

where

A', B' and Γ_1', Γ_2' are discrete points of supports in a new area and free configuration zones of membrane.

In membrane centre the maximum deflection caused by static load is determined as the sum of two deflections caused by outer load and displacement of membrane contour.

$$W_{\max} = qR^2 / 4S + 1{,}11 q\pi^3 R^2 / 24S \quad (7.2)$$

where R is the radius of circular membrane.

If membrane contour in plan is a square or a hypotrochoid fixed only in the points of contour angles, conformable mapping on the circle in the considered area can be presented as:

$$Z = W(\xi) = e^{\xi} \quad (7.3)$$

$$Z = W(\xi) = R\left(\xi + \frac{m}{\xi^n}\right) \quad |\xi| > 1 \quad (7.4)$$

$$Z = W(\xi) = R\left(\frac{1}{\xi} + m\xi^n\right) \quad |\xi| < 1 \quad (7.5)$$

R>0 is an arbitrary positive number;
n is a positive integer;
$0 \le m \le \frac{1}{n}$ is positive constant, less than one.

Hypotrochoid of z plane corresponds to the circles with radii $\rho>1$ in plane ξ.

If we insert $\xi = \frac{1}{\xi}$ into formula (7.4) we shall get the mapping of the considered area on circle $|\xi| < 1$ according to formula (7.5).

The coordinates of hypotrochoid are expressed as:

$$x = R\left(\rho\cos\theta + \frac{m}{\rho^n}\cos n\theta\right)$$
$$y = R\left(\rho sin\theta - \frac{m}{\rho^n}\sin n\theta\right) \quad (7.6)$$

Distortion factor when n=2, has the following form:

$$\frac{1}{\left|\omega'(\xi)^2\right|} = 1/R^2\left(|\xi|^2 - \frac{2\,\mathrm{Re}\,\xi^{-3}}{|\xi|^4} + \frac{1}{|\xi|^4}\right) \qquad \rho<1 \quad (7.7)$$

$$\frac{1}{\left|\omega'(\xi)^2\right|} = 1/R^2\left(1 - \frac{2\operatorname{Re}\bar{\xi}^3}{|\xi|^6} + \frac{1}{|\xi|^6}\right) \qquad \rho > 1 \tag{7.8}$$

Distortion factor when n=3 the will take the form:

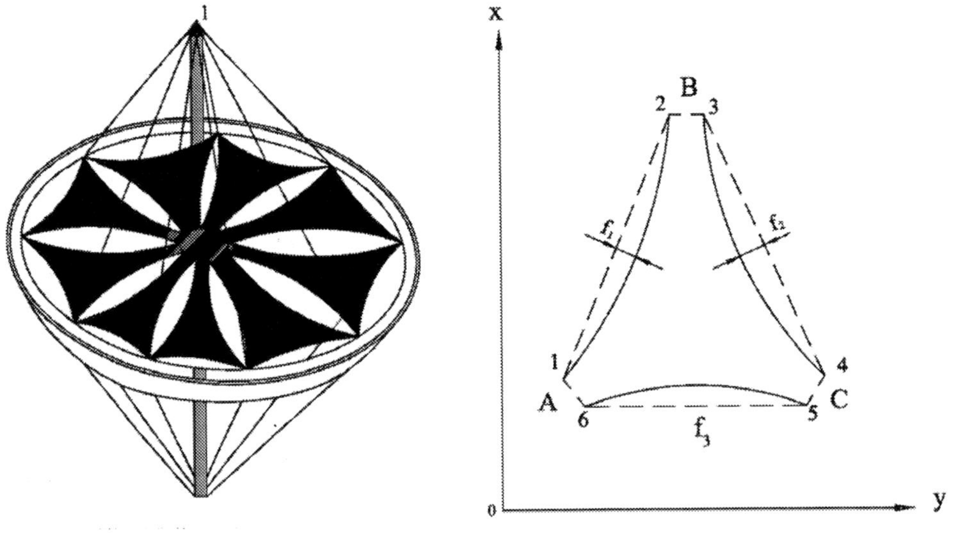

Figure 7.8. Space solar plate concentrator.

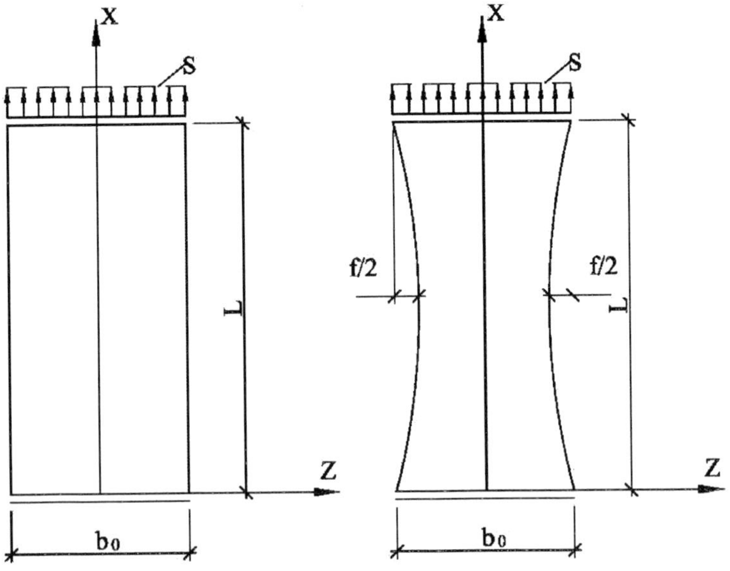

Figure 7.9. Band elements of space constructions. (Representation of membrane type)

$$\frac{1}{\left|\omega'(\xi)^2\right|} = 1/R^2\left(\left|\xi\right|^4 - \frac{2\operatorname{Re}\bar{\xi}^4}{\left|\xi\right|^4} + \frac{1}{\left|\xi\right|^4}\right) \qquad \rho<1 \tag{7.9}$$

$$\frac{1}{\left|\omega'(\xi)^2\right|} = 1/R^2\left(1 - \frac{2\operatorname{Re}\bar{\xi}^4}{\left|\xi\right|^8} + \frac{1}{\left|\xi\right|^8}\right) \qquad \rho>1 \tag{7.10}$$

If we use the above given expressions we shall get the coefficients of main natural vibrations of complex in plan configuration membranes when membrane is fixed in contour angles.

The main frequency of membrane vibrations is determined with formula:

$$\omega_{11} = \alpha\sqrt{\frac{S}{m_{red}}} \tag{7.11}$$

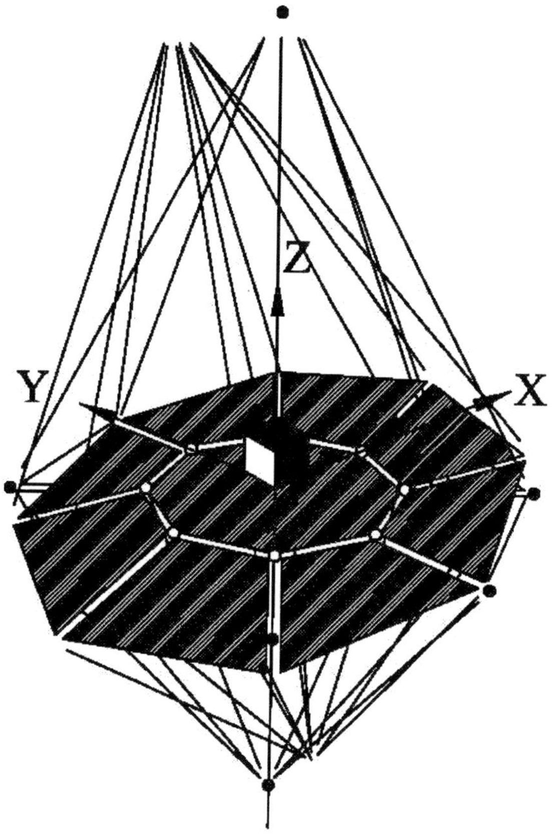

Figure 7.10. Computer modeling of space solar concentrator. (first mode of movement)

where S is membrane contour uniform tension on unit length;

m red is membrane reduced mass.

Consider as an example a curvilinear equilateral triangular membrane with side lengths ℓ=500 cm and ℓ=100000 cm. The results of calculations are given in Table which show that with the increase of boom cut the frequency of membrane natural vibrations is increased, i.e. it is possible to vary natural vibrations frequency with variation of membrane contour, as well as, with variation of membrane tension.

Comparing the obtained results we conclude that the coefficients of natural vibrations for rectangular square membrane fixed in four corners differ for 0.41% from to the results obtained by other ways.

Natural vibrations frequencies of a membrane jointly fixed on whole contour compared to membranes fixed just in separate points differ: 3.02 times for square membranes, 3.12 times for circular membranes, 3.3 times for equilateral triangular membranes.

Computer modeling was performed for space solar concentrator structure the overview, motion modes and natural strain diagrams are given in Figs. 7.8, 7.9, 7.10, 7.11, 7.12.

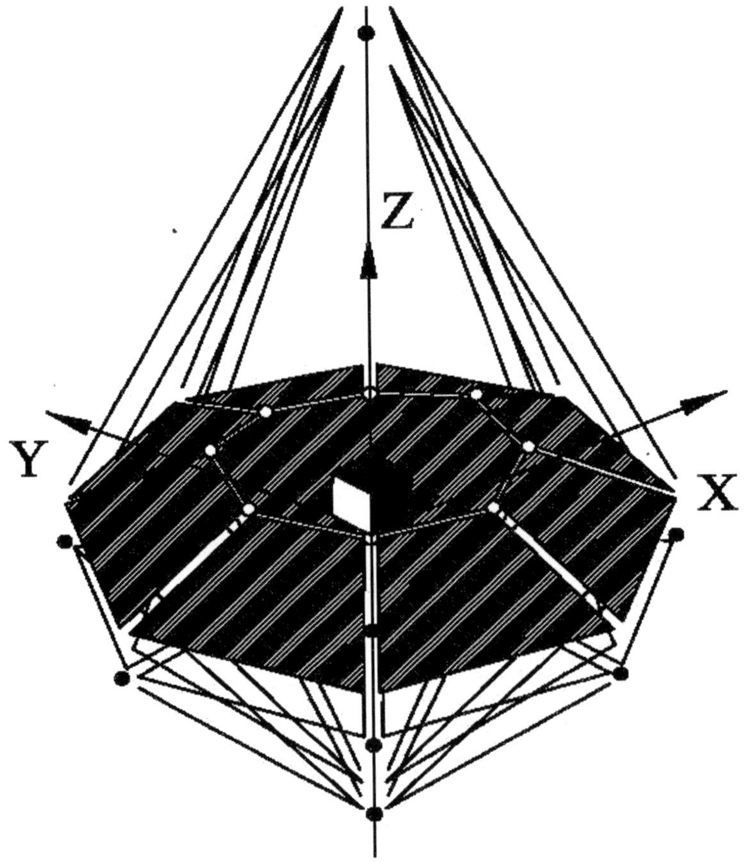

Figure 7.11. Second mode of concentrator movement.

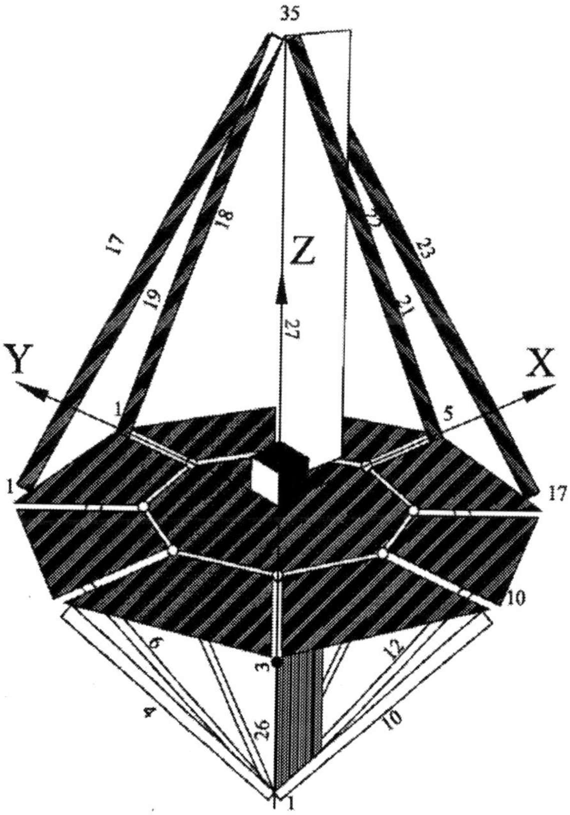

Figure 7.12. Normal strain diagrams in concentrator elements.

Coefficients of themain natural vibration frequency of the membrane of various contour in plan				
#	Outline of contour and form of fastening	Countour form of membrane	Coefficient α	Author and literature
	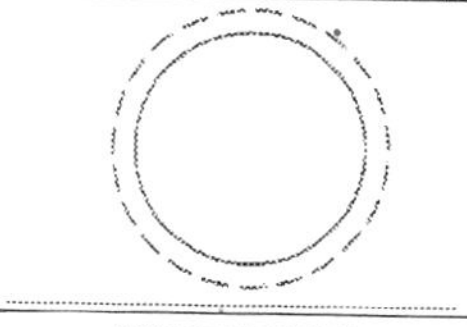	circle	4.261	Reley S.L.Timoshchenko
		square	4.443	Reley S.L.Timoshchenko
			4.43	I.M.Rabinovich et al.
	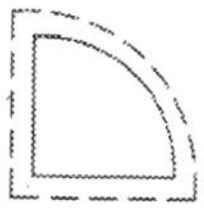	Quarter of a circle	4.551	Reley S.L.Timoshchenko

Figure 7.13. (Continued)

	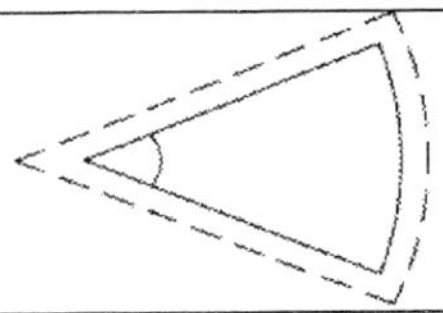	Circular section with central angle 60		4.616	“-------“
	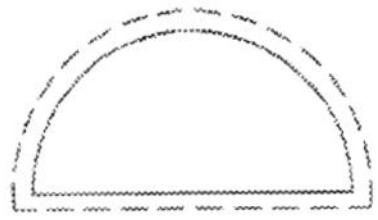	Semicircle		4.803	“-------“
		Rectangle with side ratio 2:1		4.967	“------“
		Rectangle with side ratio 3:1		5.736	“------“
		Reclangle with side ration 3:2		4.624	“-------“
	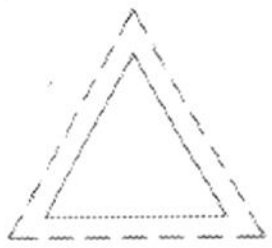	Equilateral triangle		4.774	Releу S.L.Timoshchenko
				4.797	B.G.Korenev
	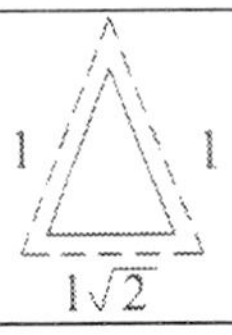	Isosceles triangle		4.967	Releу S.L.Timoshchenko
		Circle		1.367	I.M. Rabinovich et al.
		Square		1.469	I.M.Rabinovich et al.
				1.463	
	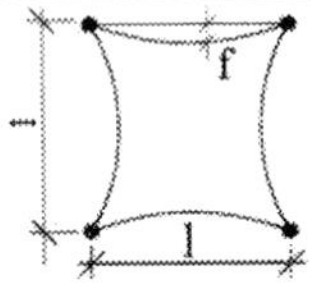	Curvilinear square	f/l=0 f/l=0.08 f/l=0.16 f/l=0.26	1.452 1.467 1.476 1.479	Autores
	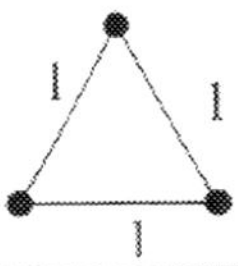	Equilateral triangle		1.447	Autores

	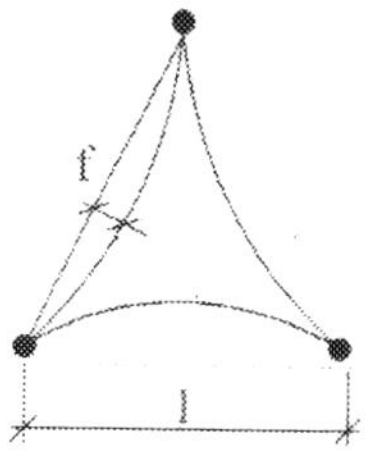	Curvilinear equilateral triangle	f/l=0.05 f/l=0.1 f/l=0.15	1.461 1.471 1.476	Autores

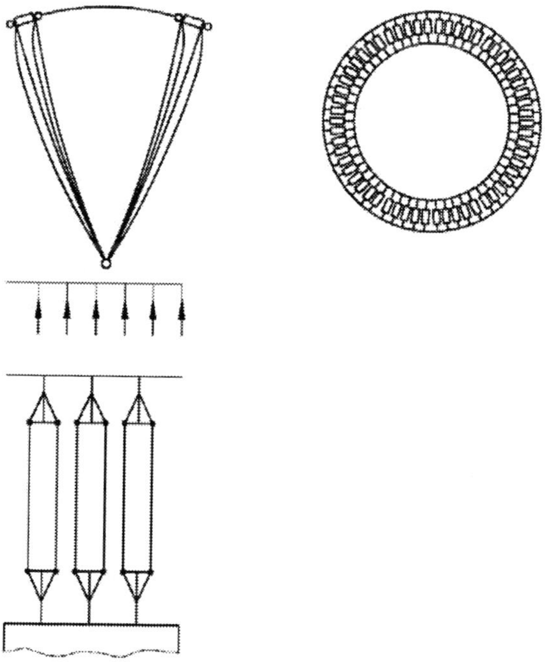

Figure 7.13. Open-bend constructions.

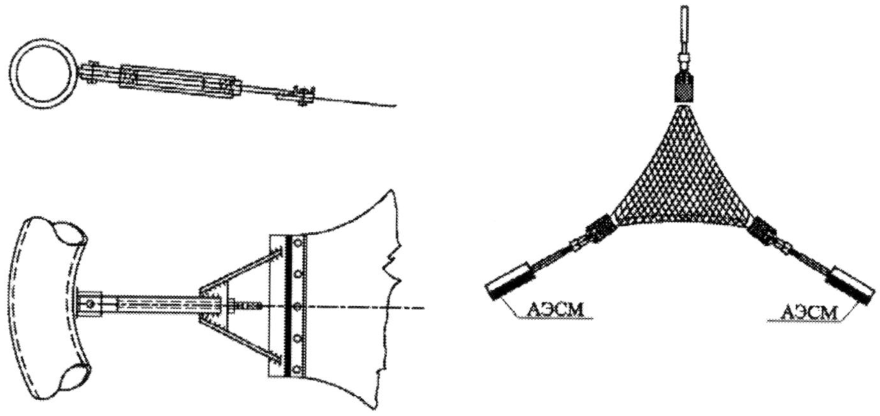

Figure 7.14. Scheme of bend tension for a spece bend.

7.3. Regulation of Vibration Frequency of Variable Width Band Membranes

Constant and variable width band membranes are also used in rotating reflecting mirrors (reflectors).

If we consider band membrane as zero-moment shell-membrane, its stress-strained state is expressed with formula:

$$S_x \frac{\partial^2 W}{\partial x^2} + 2T_{xy} \frac{\partial^2 W}{\partial x \partial y} + S_y \frac{\partial^2 W}{\partial y^2} = -q\sqrt{1 + \left(\frac{\partial W}{\partial x}\right)^2 + \left(\frac{\partial W}{\partial y}\right)^2} \qquad (7.12)$$

Here:

S_x and S_y are longitudinal stresses per unit length;

T_{xy} is shift stress per unit length;

q is load intensity per unit area.

If we assume that the membrane is long and inclined enough then the right part of equation (7.12) can be taken as equal to q. In this case error to membrane contour is approximately:

$f/\ell = \frac{1}{20} = 0,05$ approximately 2%;

$f/\ell = \frac{1}{15} = 0,07$ approximately 3.5%;

$f/\ell = \frac{1}{12} = 0,09$ approximately 5%;

$f/\ell = \frac{1}{10} = 0,1$ approximately 8%;

According to Rhamaswami state the conditions:

$T_{xy}=0$; $S_y=0$; and $S_x=S=const$

Then equation (7.12) takes the form:

$$\frac{d^2W}{dx^2} = -\frac{q(x)}{S} \qquad (7.13)$$

Denote $\rho=\gamma/g$; γ is volume weight of material; g is free fall acceleration; ℓ is length; b_0 is band membrane width; $b_0S=H$.

Using d'Alembert's principle we shall have:

$$\rho(x)\frac{\partial^2 W}{\partial t^2}=\frac{H}{b_0}\frac{\partial^2 W}{\partial x^2} \tag{7.14}$$

Boundary conditions: W=0 when x=0 and x=ℓ;
Accept the law of band membrane width variation as (Figure 7.9):

$$F(x)=F_0\left[1+\frac{16f^2}{\ell^2}\left(1-2\frac{x}{\ell}\right)^2\right]^{1/2} \tag{7.15}$$

Here we have $F_0 = b_0\delta$ (δ is membrane thickness).

Use equation (7.14) for solution. Considering (7.15) we get main frequencies of natural vibrations of band membrane when membrane width changes as:

$$\omega_1=\alpha\frac{1}{\ell}\sqrt{\frac{H}{\rho F_0}} \tag{7.16}$$

Here I, II and III are the number of members of expansion into series considered in the solution.

Numerical tests carried out on circular composite membrane with program "SAP-2000 student" gave the results that show that with variation of guy system tension the compliance in membrane fastening points are changed that by itself affects the period of guy-membrane system vibrations. Thus, for example, the 10 times decrease of compliance in membrane fix points changes oscillation period from 0.5205 sec to 1.7078 sec, i.e. increases 3.3 times; the increase of the same compliance for 10 times causes the decrease of oscillation period from 0.5205 sec to 0.1815 sec, i.e. the decrease for 2.9 times.

Table 7.2.

Frequency of natural oscillations of variable width band membrane					
		ω_1 1/sec			Constant width band membrane ω_1 1/sec
f/ℓ	$H=S/b_0$	I	II	III	
0,17	0,09	9,075	8,973	8,969	8,603
0,25	0,06	7,472	7,251	7,237	7,025
0,5	0,03	5,485	5,046	4,998	4,967

Chapter 8

PROSPECTS OF THE DEVELOPMENT OF CONTROLLED STRUCTURES

1. SOME PROBLEMS OF PERSPECTIVE REGULATION OF TENSIONS

Metal structures of the nearest future will be the structures of large span and complicated constructions the realization of which in common metal is uneconomical and sometimes even impossible. Wide application will be acquired by prestressed constructions: suspended, guy, spatial, of aluminum alloys, combined (combined of different materials with the use of different high-strength materials and plastics). Concentration of material, application of stressed stretched surfaces of single and double curvature will be developed in new constructions.

The up-dating of constructive form of industrial buildings and structures should be realized in the following basic directions:

a) Development of new constructions using the method of prestressing enabling to regulate forces in construction members;
b) Using on continuous and spatial systems;
c) Creation of combined constructions using high-strength cables, orthotropic metal platings, reinforced concrete, etc.

In order to tension regulate in the future, i.e. at updating, reconstruction and other similar cases the most optimum are the following constructions and structures:

1. Constructions where the frame is possibly less depended on the change and updating of the technological process. In this respect the cross-section diagram presented in Figure 8.1 is very characteristic, where for the convenience of redesign of the plant in the future (widening of the building, increase of exploitation capacity) the cross-section diagram of the building is taken as statically definable. Here one may consider workshop buildings with floor transport or crane separate type columns having separate crane branches.
2. Constructions where maximum independence of main bearing constructions from constructions supporting the building enclosure is considered. For example, crane

platform in the middle row of columns of open-hearth plant, constructed independently from the rest of the framework can be easily updated and in case of necessity, strengthened. Such decision was made for open-hearth plant with furnaces of 250-500 tons.

3. Constructions where tension systems are used, the characteristic feature of them being the reliability at overloads. For example, in guy systems tension can be easily regulated.
4. Metal constructions are thin walled by the principle of their formation. Also, rational rolling sections used in constructions make easier the task of stress regulation.

The indicator of the advisability of steel used in cross-section of, for example, central compressed rod, is the so called coefficient of form. mode

$$K_f = \frac{r_{\min}^2}{F}$$

where r_{min} is the least radius of cross-section inertia;

F is the area of cross-section.

For the section consisting of two angles:

with T arrangement K_f=0,18÷0,27

for U-bar section K_f=0,14÷0,18

for I-shaped section K_f=0,09÷0,11

for tubular section K_f=0,5÷1,5

Thus, at equal consumption of steel the flexibility λ of tubular section is 2-6 times less than the flexibility of usually used section of two angles while bearing capacity of tubular section is considerably more. In tubular section it is most opportune to arrange tension elements – cables, in order to create prestressing in stretched members of constructions.

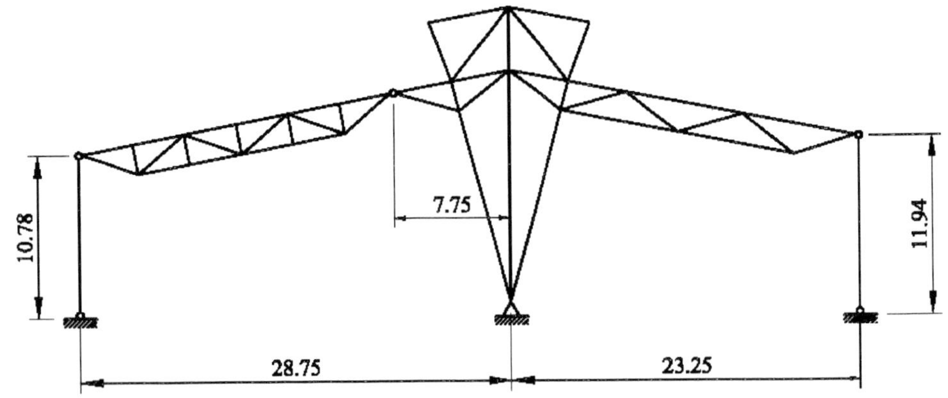

Figure 8.1. Cross-section diagram of steel welding shop structure in Ruorkerla.

The use of formed sections can simplify the task of tension regulation at construction strengthening as formed sections can be given practically any required shape providing the necessary flexibility of the rod. At present the formed sections are made of steel band on roll-bending benches.

In the designed constructions for the purpose of prospective regulation of forces (tensions) in separate cases the possibility of changing the construction diagram in the future by: a) restraint of support devices; b) introduction of braces; c) use of tightening devices, for example, strut frame; d) tightening of constructions; e) underpinning of additional intermediate supports can be considered.

The installation of additional joints or rivets or gluing of additional members can also be provided beforehand.

Such are in the brief the main ways of development of future regulation of tensions.

INDEX

A

B

C

M

N

O

P

Q

R

S

T

U

V

W

Y

Z